Agentic AI in Manufacturing & Industry 4.0

Practical Design Patterns for Building Autonomous Intelligent Systems Across Industries

A Comprehensive Hands-On Guide to Agentic AI Frameworks, Multi-Agent Reasoning, and Scalable Autonomous Workflows

Dr. Tayeb Ayat

ISBN 979-8-90604-007-7

Acknowledgments

No words can fully capture the gratitude I feel toward my family. To my wife Minoo — your love has carried me through every challenge and made every achievement meaningful. To Elvin and Ella — you are my greatest teachers and my deepest joy. I write so that you might one day understand the world a little better, and so that this work might honor the home you have given me.

Contents

How to Use This Book

This book is designed as a progressive, hands-on journey into the architecture of autonomous AI systems. Whether you are an AI engineer looking to move beyond simple prompt chaining, a data scientist deploying reasoning systems, or a technical leader orchestrating cross-functional AI teams, this guide provides the blueprints you need to build reliable, maintainable, and scalable agentic workflows.

You can read this book sequentially to build your understanding from foundational concepts to advanced multi-agent ecosystems, or you can use it as a desk reference, jumping directly to specific Agentic Design Patterns as you encounter architectural challenges in your own projects. The progressive layers presented here form a complete agentic architecture: one that senses, reasons, plans, and acts autonomously.

Each chapter follows a strict four-step structure to take you from concept to deployment. We begin with a **Pattern Overview** that explains the cognitive and architectural principles behind the pattern. Next, we explore **Industry Use Cases** to see how real companies apply this pattern to drive impact. We then dive into **Hands-On Implementation** with end-to-end examples built using leading frameworks like LangChain, CrewAI, and Google's Agent Development Kit. Finally, we conclude with **Key Takeaways**, covering lessons learned, performance metrics, and ways to extend the pattern.

Code examples are provided throughout, focusing on practical, deployable solutions. We highly recommend setting up a dedicated development environment to run the examples, experiment with the parameters, and adapt the patterns to your specific industry needs. The skills you learn here are designed to be transferable and framework-agnostic, preparing you for any ecosystem.

Preface

Artificial Intelligence is undergoing a fundamental shift. We are moving away from reactive models that simply answer prompts, toward agentic systems—entities capable of perceiving their environment, reasoning through complex problems, orchestrating workflows, and taking autonomous action to achieve long-term goals with minimal human supervision. *Agentic AI in Manufacturing & Industry 4.0* is your comprehensive guide to navigating and mastering this transition.

This book bridges the gap between theoretical cognitive architectures and production-ready engineering. It provides a robust library of Agentic Design Patterns—battle-tested templates that capture core capabilities such as contextual reasoning, dynamic planning, tool invocation, self-reflection, and multi-agent collaboration. By mastering these patterns, you will gain the ability to engineer systems that adapt, communicate, and improve over time, functioning much like intelligent human collaborators.

Throughout these pages, we explore implementations across today's leading frameworks, including LangChain, CrewAI, n8n, and the Microsoft Agent Framework. Furthermore, a dedicated portion of this book dives deep into the Google Gemini Enterprise Agentic Ecosystem, offering a production-grade perspective on deploying secure, scalable, and observable autonomous workflows using the latest Gemini models and Google Cloud infrastructure.

Agentic AI is not merely a trend; it is the foundational architecture of the next decade of software. As industries integrate intelligent agents into logistics, operations, customer service, and analytics, the ability to design agentic systems will become as essential as knowing how to code. By the end of this book, you will be equipped to design and deploy intelligent systems that do more than predict—they decide, act, and collaborate.

About the Author

Dr. Tayeb Ayat

Data Science & Agentic AI Solutions Leader

Dr. Tayeb Ayat builds high-performing data science teams that deliver measurable business transformation. Over more than twelve years, he has scaled AI/ML organizations from the ground up across retail and energy operations — achieving exceptional retention while driving enterprise-wide adoption of cutting-edge technologies.

What drives him is the intersection of AI innovation and real-world business outcomes: building organizations that deploy autonomous systems at scale, not just models in notebooks. He combines deep technical expertise in the latest AI — agentic systems, multi-agent orchestration, generative AI, computer vision — with proven executive leadership: securing C-suite investment, negotiating enterprise partnerships with major cloud providers, and mobilizing large teams through strategic change management.

He has led platform consolidations, built data governance frameworks, and pioneered emerging AI capabilities in production — and built organizations where exceptional talent thrives and stays. His philosophy is simple: start with the business problem, build the team that can solve it, implement the technology that scales it, and measure the impact that matters.

He is currently exploring the evolution of agentic AI at enterprise scale, multi-agent orchestration patterns, and how the Model Context Protocol (MCP) will reshape intelligent systems.

SECTION

Manufacturing & Industry 4.0

This section explores the transformative impact of autonomous intelligent systems on modern production environments, detailing how agentic AI bridges the gap between physical operations and digital intelligence. Readers will dive deep into practical design patterns for orchestrating **Industrial Robotics & Cells**, optimizing **Supply Chain & Logistics**, and deploying edge-native agents for real-time **Quality Assurance & Inspection**. Expect rigorous architectural blueprints and hands-on workflows designed to scale resilient, connected factories.

A Note on Model Versions & Staying Current

This book was written at a fast-moving moment in artificial intelligence. The model versions it names—principally Google's **Gemini 3.5** family, alongside **OpenAI GPT-5.5**, **Anthropic Claude** (Opus 4.8 / Fable 5), and **Meta Llama 4**[*]—reflect the state of the art in mid-2026. By the time you read this, newer models, SDK versions, and product names will almost certainly exist. *That is by design:* the lasting value of this book is its **architecture and patterns**, which outlive any single model release.

So as you read:

- **Always use the latest available model.** Where the text names a specific version, substitute your provider's current flagship; the pattern is unchanged.
- **Expect API and SDK syntax to drift.** Method names, parameters, and import paths evolve between releases. Before running any code listing, check it against the current official documentation and confirm it compiles in your environment.
- **Product names change.** Platforms get renamed and consolidated; focus on the capability being described, not the brand on it.

The concepts are durable; the version numbers are a snapshot. Build on the first, and verify the second.

[*]Model and product versions cited throughout this book reflect the state of the art in mid-2026 and *will* be superseded. Wherever a specific version appears, read it as "the best model available to you right now" and substitute your provider's current flagship—the architectural patterns are deliberately model-agnostic.

Chapter 1

Manufacturing Operations

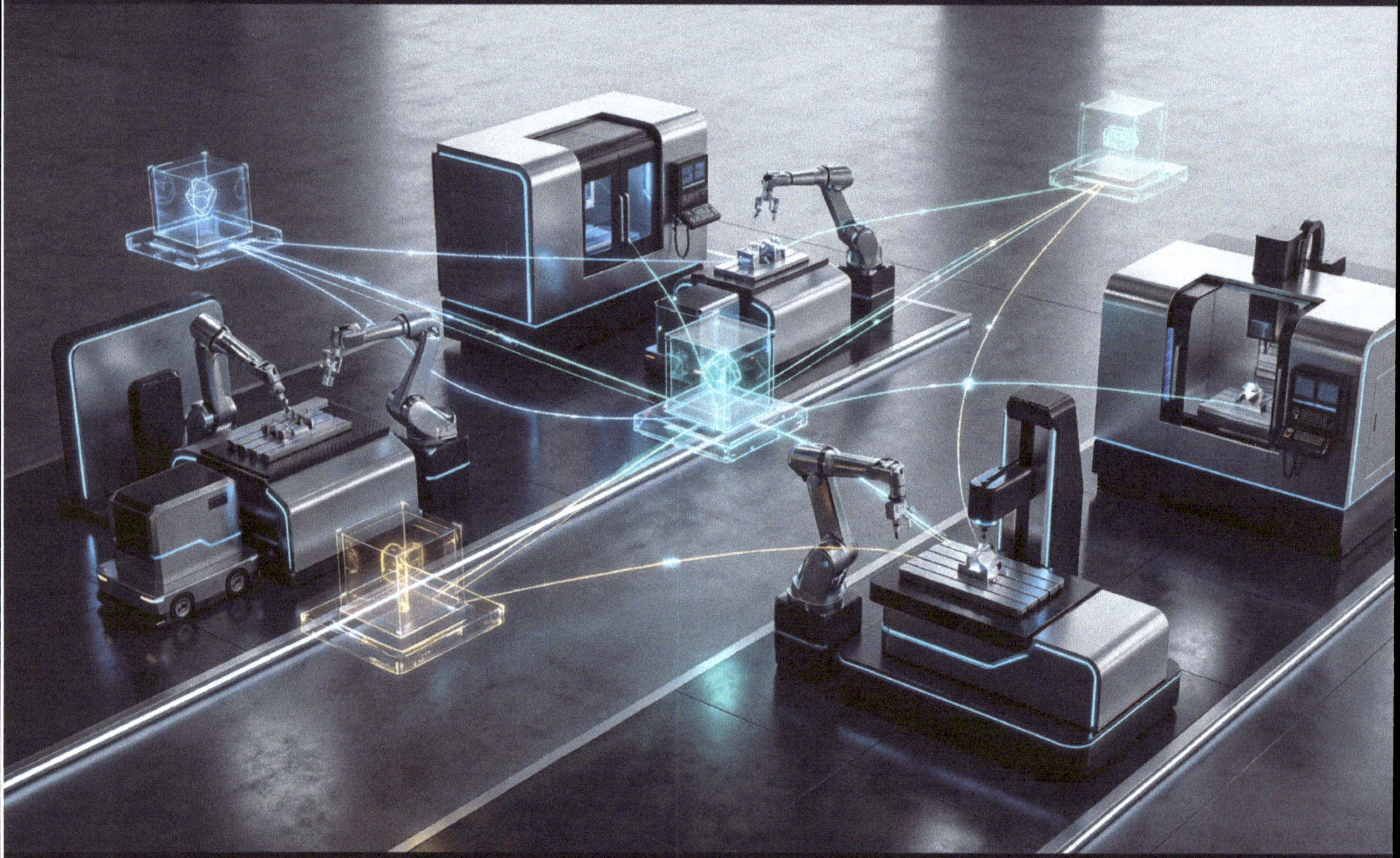

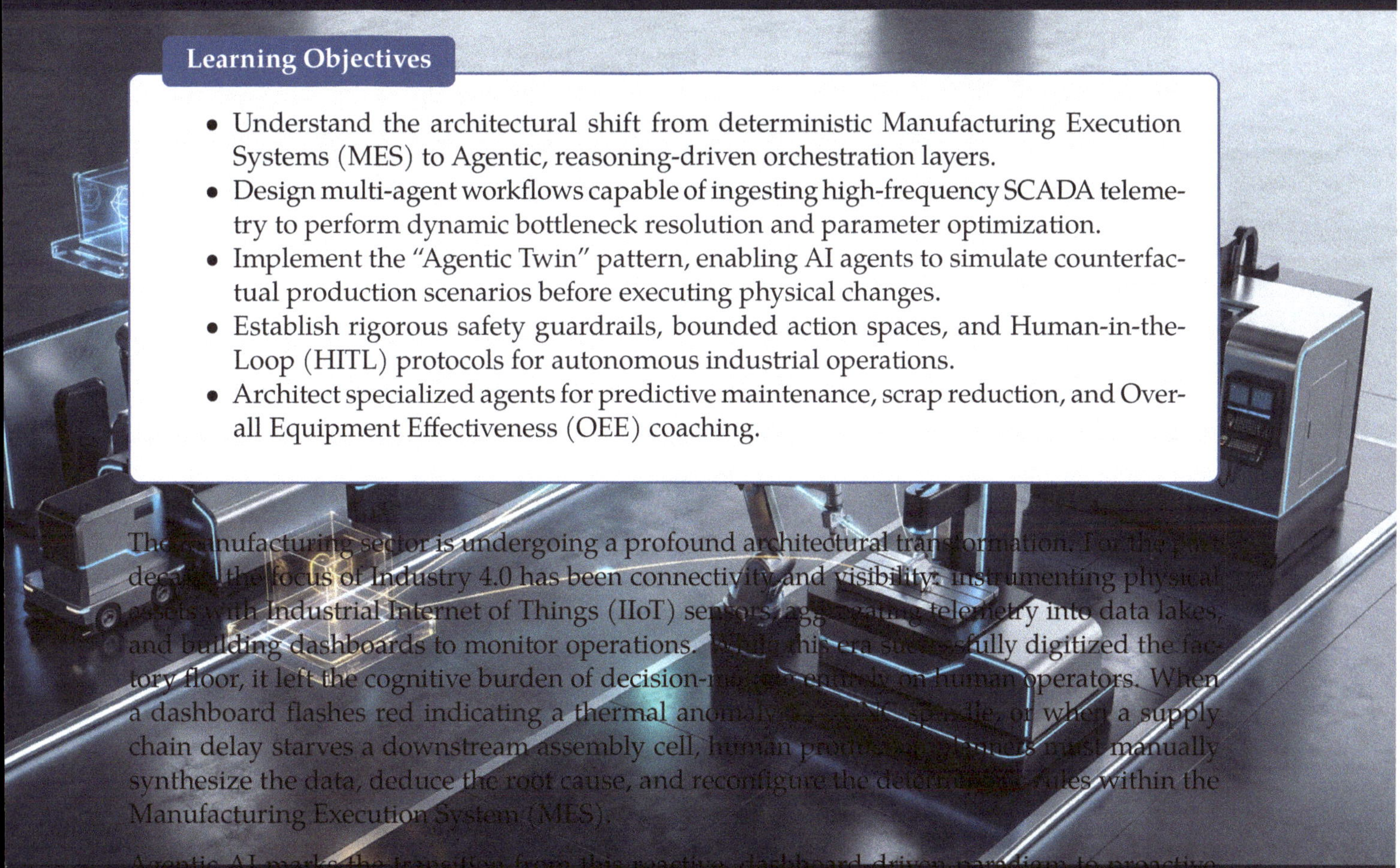

Learning Objectives

- Understand the architectural shift from deterministic Manufacturing Execution Systems (MES) to Agentic, reasoning-driven orchestration layers.
- Design multi-agent workflows capable of ingesting high-frequency SCADA telemetry to perform dynamic bottleneck resolution and parameter optimization.
- Implement the "Agentic Twin" pattern, enabling AI agents to simulate counterfactual production scenarios before executing physical changes.
- Establish rigorous safety guardrails, bounded action spaces, and Human-in-the-Loop (HITL) protocols for autonomous industrial operations.
- Architect specialized agents for predictive maintenance, scrap reduction, and Overall Equipment Effectiveness (OEE) coaching.

The manufacturing sector is undergoing a profound architectural transformation. For the past decade, the focus of Industry 4.0 has been connectivity and visibility: instrumenting physical assets with Industrial Internet of Things (IIoT) sensors, aggregating telemetry into data lakes, and building dashboards to monitor operations. While this era successfully digitized the factory floor, it left the cognitive burden of decision-making entirely on human operators. When a dashboard flashes red indicating a thermal anomaly in a CNC spindle, or when a supply chain delay starves a downstream assembly cell, human production planners must manually synthesize the data, deduce the root cause, and reconfigure the deterministic rules within the Manufacturing Execution System (MES).

Agentic AI marks the transition from this reactive, dashboard-driven paradigm to proactive, [illegible]

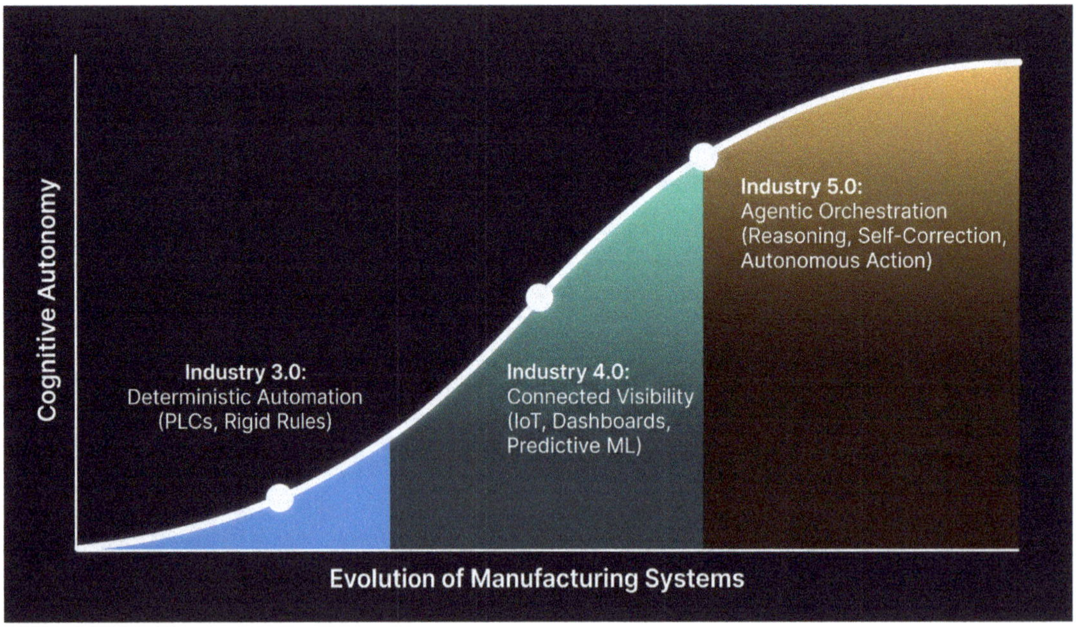

Figure 1.1: The maturity curve of manufacturing systems, illustrating the shift from deterministic automation to cognitive, agentic orchestration.

To understand why agentic architectures are necessary, we must examine the limitations of traditional manufacturing software. Traditional MES and Enterprise Resource Planning (ERP) systems operate on rigid, deterministic rule engines. They excel at executing predefined sequences (e.g., "If Part A is scanned, route to Station B"). However, the physical world is inherently non-deterministic. Tool wear degrades non-linearly, ambient humidity affects material curing times, and micro-stoppages cascade into macro-level bottlenecks. When a novel exception occurs—one that was not explicitly programmed into the PLC (Programmable Logic Controller) or MES logic—the system halts and waits for human intervention.

Agentic systems bridge this gap by introducing semantic reasoning. An agent can ingest a stream of error codes, cross-reference them with historical maintenance logs stored in a vector database, query the digital twin for the current physical state of the machine, and dynamically generate a novel recovery sequence. This capability transforms the concept of the "Digital Twin" from a passive, read-only mirror of physical assets into an *Agentic Twin*—an active entity capable of running counterfactual simulations ("If I reduce the feed rate by 5%, will it prevent the thermal trip without violating the shift's throughput quota?") and acting upon the optimal result.

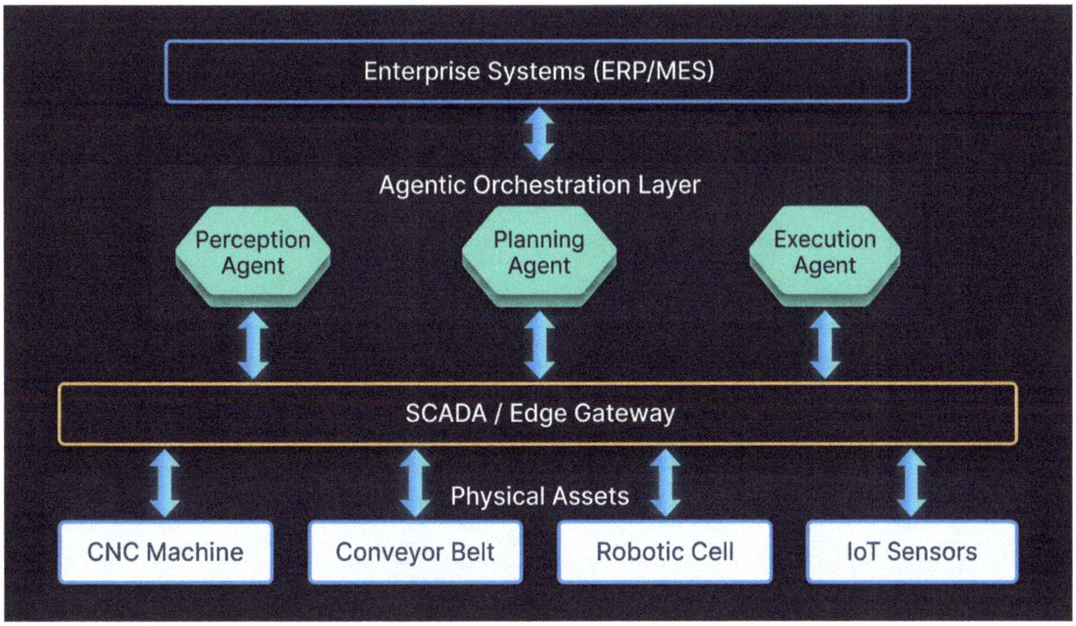

Figure 1.2: Layered architecture of an Agentic Manufacturing System, demonstrating how cognitive agents sit between enterprise planning systems and physical execution layers.

Implementing agents in a manufacturing context requires strict adherence to safety and determinism. Unlike a customer service chatbot where a hallucination might result in a confused user, a hallucination on a factory floor could result in a robotic arm colliding with a fixture, causing hundreds of thousands of dollars in damage or risking human safety. Therefore, manufacturing agents are architected using the *Planner-Executor-Evaluator* pattern with rigid, API-enforced boundaries.

The agentic layer does not replace the deterministic control layer (the PLCs and safety relays); rather, it sits above it. The agent acts as a highly intelligent supervisor that can only invoke specific, pre-approved tools. For example, an agent cannot arbitrarily command a motor to spin at 10,000 RPM. Instead, it calls a tool like `adjust_spindle_speed(target_rpm=8000)`, which is intercepted by an edge gateway that validates the request against hard-coded safety limits before passing it to the PLC. Furthermore, for high-stakes decisions, the architecture mandates a Human-in-the-Loop (HITL) gateway, where the agent prepares a comprehensive briefing of the anomaly, the proposed solution, and the simulated outcome, presenting it to a human engineer for one-click approval.

To illustrate the foundational structure of such a system, consider the following LangGraph state definition for a factory floor orchestrator. This code defines the state schema that flows between specialized agents (Perception, Planning, and Execution) as they collaboratively resolve a production bottleneck.

Listing 1.1: Defining the state schema for a Multi-Agent Factory Orchestrator using LangGraph

```python
from typing import TypedDict, Annotated, List, Dict, Any
from langgraph.graph import StateGraph, END
import operator

```

```python
# Define the state that persists across the agentic workflow
class FactoryState(TypedDict):
    # Raw telemetry ingested from the SCADA/IoT gateway
    telemetry_data: Dict[str, Any]

    # Identified issues (e.g., "Thermal anomaly on CNC_04")
    active_alarms: List[str]
    current_bottleneck: str

    # The agent's proposed intervention
    proposed_action: Dict[str, Any]

    # Simulated impact on Overall Equipment Effectiveness (OEE)
    oee_impact_score: float

    # Safety validation and execution status
    safety_clearance: bool
    execution_status: str

# Initialize the state graph
workflow = StateGraph(FactoryState)

# Node 1: Perception Agent (Ingests data, identifies anomalies)
def perception_node(state: FactoryState) -> FactoryState:
    # Logic to query SCADA APIs and update active_alarms
    pass

# Node 2: Planning Agent (Generates parameter adjustments)
def planning_node(state: FactoryState) -> FactoryState:
    # Logic to reason over alarms and propose an action
    pass

# Node 3: Safety & HITL Gateway (Validates action against constraints)
def safety_gateway_node(state: FactoryState) -> FactoryState:
    # Logic to check bounds or request human approval
    pass

# Node 4: Execution Agent (Dispatches commands to MES/Edge)
def execution_node(state: FactoryState) -> FactoryState:
    # Logic to invoke physical actuation tools
    pass
```

The ultimate goal of deploying these agentic architectures is the optimization of Overall Equipment Effectiveness (OEE)—the gold standard metric in manufacturing, calculated as the product of Availability, Performance, and Quality. Traditional optimization efforts treat these three variables in isolation. An agentic system, however, understands the complex trade-offs between them. It knows that pushing a machine to run 10% faster (increasing Performance) might increase vibration, leading to a 2% increase in micro-fractures (decreasing Quality) and accelerating tool wear (decreasing future Availability). By reasoning over these multi-dimensional trade-offs in real-time, agents can continuously steer the factory toward the global optimum.

What This Chapter Covers

In the following sections, we will dive deep into the specific agentic design patterns and specialized roles required to run an autonomous manufacturing operation. We will move beyond the theoretical architecture and explore concrete, production-ready implementations for the most critical factory workflows.

- **Predictive Maintenance:** How agents transition maintenance from scheduled intervals to condition-based, autonomous interventions by reasoning over vibration, thermal, and acoustic telemetry.
- **Throughput & Bottleneck Planner:** Architecting agents that dynamically re-route Work-in-Progress (WIP) and adjust line speeds to resolve cascading bottlenecks in real-time.
- **Recipe/Parameter Optimizer:** Utilizing agents to continuously fine-tune machine parameters (feed rates, temperatures, pressures) to account for environmental drift and raw material variance.
- **Scrap & Rework Root-Cause Agent:** Designing investigative agents that trace product defects back through the manufacturing genealogy to identify and eliminate the root causes of scrap.
- **Energy-per-Unit Minimizer:** Implementing agents that optimize the energy consumption of heavy industrial equipment without compromising production quotas.
- **OEE Coach:** Building an overarching meta-agent that synthesizes insights from all sub-agents to provide strategic, natural-language guidance to plant managers for continuous improvement.

Each section will detail the specific tool schemas, prompt engineering techniques, and safety guardrails required to deploy these agents securely on the factory floor, leveraging frameworks like Google's Agent Development Kit (ADK) and LangGraph.

CHAPTER 260

Predictive Maintenance

From Condition Monitoring to Autonomous Asset Orchestration

◎ Learning Objectives

- Understand the architectural evolution from traditional Machine Learning anomaly detection to Agentic Predictive Maintenance.
- Design multi-agent workflows that ingest multi-modal telemetry (vibration, thermal, acoustic) and translate it into semantic context for LLM reasoning.
- Implement the "Agentic Twin" pattern to allow agents to simulate the production impact of maintenance interventions before execution.
- Architect specialized agent roles (Perception, Diagnostic, Planning, Execution) using frameworks like LangGraph.
- Establish rigorous safety guardrails and Human-in-the-Loop (HITL) approval gateways for autonomous industrial control.

1.1 The Evolution of Maintenance: From Heuristics to Agentic Reasoning

The discipline of industrial maintenance has undergone several paradigm shifts over the past century, each driven by advancements in data collection and computational power. To understand the profound impact of Agentic AI on this field, we must first trace this evolutionary trajectory and identify the critical limitations of the systems that preceded it.

Historically, manufacturing facilities operated on a *Reactive* or "run-to-failure" model. Assets were operated until they broke down, at which point production halted, and emergency repairs were initiated. The economic cost of this approach—measured in unplanned downtime, expedited shipping for spare parts, and catastrophic secondary damage to machinery—was astronomical. This gave way to *Preventive Maintenance* (PM), a time-based or cycle-based heuristic approach. Engineers consulted Original Equipment Manufacturer (OEM) manuals and scheduled bearing replacements every 10,000 hours of operation, regardless of the actual condition of the bearing. While this reduced catastrophic failures, it introduced massive inefficiencies: perfectly healthy components were discarded prematurely, and maintenance labor was wasted on unnecessary teardowns.

The advent of Industry 4.0 introduced *Condition-Based Monitoring* (CBM) and traditional *Predictive Maintenance* (PdM). By instrumenting assets with Industrial Internet of Things (IIoT) sensors, facilities could monitor vibration, temperature, and acoustic emissions in real-time. Data scientists deployed traditional Machine Learning models—such as Random Forests, Support Vector Machines, and Long Short-Term Memory (LSTM) networks—to detect anomalies and calculate the Remaining Useful Life (RUL) of a component.

However, traditional ML-based Predictive Maintenance suffers from a critical limitation: it lacks *agency* and *contextual reasoning*. An LSTM model might successfully predict that a CNC spindle bearing will fail in 48 hours with 90% confidence. It outputs this prediction to a dashboard, turning a digital indicator red. But the model stops there. It does not know if the required replacement bearing is in the warehouse. It does not know that the CNC machine is currently executing a critical, high-margin aerospace order that must ship tomorrow. It cannot read the OEM manual to determine the standard repair time, nor can it generate a work order in the Enterprise Resource Planning (ERP) system. The cognitive burden of synthesizing the prediction, the supply chain reality, and the production schedule falls entirely on human reliability engineers and production planners.

The Agentic Paradigm Shift

Agentic Predictive Maintenance bridges this gap by moving from isolated predictive models to interconnected reasoning systems. In an agentic architecture, the anomaly detection model is merely a *tool*—a sensory input—utilized by a higher-order cognitive agent powered by a Large Language Model (LLM).

When an agentic system detects the same spindle bearing degradation, it does not merely flash a red light. The system initiates a multi-step reasoning and execution workflow: 1. **Contextualization:** The agent queries the SCADA system to confirm the anomaly and retrieves the specific error codes. 2. **Diagnosis:** It uses Retrieval-Augmented Generation (RAG) to search historical maintenance logs and OEM manuals, determining that this specific vibration signature at 12,000 RPM indicates inner-race spalling. 3. **Supply Chain Verification:** It autonomously queries the ERP system via API to check the inventory for the specific bearing part number. 4. **Production Planning:** It queries the Manufacturing Execution System (MES) to view the current production schedule, identifying a 4-hour planned changeover window occurring in 36 hours. 5. **Execution & Orchestration:** It drafts a comprehensive work order, reserves the part in the ERP, schedules a technician for the changeover window, and—crucially—can dynamically adjust the machine's parameters (e.g., reducing the feed rate by 10%) to ensure the bearing survives until the scheduled maintenance window, thereby saving the current production run.

This represents a fundamental shift from deterministic, reactive software to semantic, proactive orchestration. The agent understands the multi-dimensional trade-offs between asset health, supply chain constraints, and production quotas. By automating the cognitive heavy lifting of maintenance planning, agentic systems drastically reduce Mean Time to Repair (MTTR), eliminate the silos between Operational Technology (OT) and Information Technology (IT), and maximize Overall Equipment Effectiveness (OEE).

1.2 Architecting the Predictive Maintenance Multi-Agent System

Building an agentic predictive maintenance system requires a robust, distributed architecture. A single monolithic LLM prompt cannot reliably handle the complexities of real-time telemetry ingestion, RAG-based diagnostics, ERP integration, and SCADA actuation. Instead, we employ a Multi-Agent System (MAS) architecture, where specialized agents with distinct personas, toolsets, and bounded contexts collaborate to resolve maintenance events.

Using frameworks like LangGraph, we can structure this workflow as a stateful, cyclic graph. The architecture typically consists of four primary agent roles: The Perception Agent, the Diagnostic Agent, the Planning Agent, and the Execution Agent.

1.2.1 The Agent Roles and Responsibilities

1. The Perception Agent (The Watcher): This agent operates at the edge or near-edge. Its primary responsibility is to monitor the continuous stream of telemetry data from the SCADA system or IoT gateway. It does not process raw high-frequency data directly (as discussed in Section 3); rather, it monitors the outputs of edge-deployed ML anomaly detection models. When an anomaly threshold is breached, the Perception Agent gathers the surrounding context—current machine state, active recipe, ambient conditions—and formats this into a structured semantic alert to initiate the workflow.

2. The Diagnostic Agent (The Investigator): Triggered by the Perception Agent, the Diagnostic Agent is responsible for root-cause analysis. It is equipped with RAG tools connected to a vector database containing OEM manuals, Failure Mode and Effects Analysis (FMEA) documents, and historical maintenance logs. Its goal is to translate a symptom (e.g., "High vibration on Axis Z") into a specific diagnosis (e.g., "Ball screw degradation likely due to insufficient lubrication").

3. The Planning Agent (The Scheduler): Once a diagnosis is established, the Planning Agent takes over. This agent is heavily integrated with enterprise systems. It uses API tools to query the ERP for spare parts inventory and the MES for the production schedule. It must solve a complex optimization problem: finding the optimal time to perform the maintenance that minimizes disruption to high-priority orders while ensuring the machine does not fail catastrophically in the interim.

4. The Execution Agent (The Dispatcher): The final node in the graph is the Execution Agent. Based on the plan, it executes the necessary API calls to generate a work order, reserve parts, and notify technicians. In advanced, closed-loop systems, the Execution Agent may also call SCADA APIs to adjust machine parameters (e.g., reducing speed) to extend the asset's life until the scheduled repair window.

1.2.2 Implementation with LangGraph

To implement this architecture, we define a shared state object that persists across the agentic workflow. This state acts as the "whiteboard" where agents read inputs and write their findings. Below is a production-grade LangGraph implementation demonstrating this stateful orchestration.

Listing 1.2: Defining the Multi-Agent Predictive Maintenance Workflow using LangGraph

```python
from typing import TypedDict, Annotated, List, Dict, Any
from langgraph.graph import StateGraph, END
from langchain_core.messages import BaseMessage, HumanMessage, AIMessage
import operator

# 1. Define the shared state schema
class MaintenanceState(TypedDict):
    asset_id: str
    raw_telemetry_alert: Dict[str, Any]
    semantic_symptoms: List[str]
    diagnosis: str
    confidence_score: float
    required_parts: List[str]
    parts_available: bool
    proposed_schedule: str
    execution_plan: Dict[str, Any]
    hitl_approved: bool
    messages: Annotated[List[BaseMessage], operator.add]
```

```python

# 2. Initialize the StateGraph
workflow = StateGraph(MaintenanceState)

# 3. Define Agent Nodes
def perception_node(state: MaintenanceState) -> MaintenanceState:
    """Translates raw alerts into semantic symptoms."""
    alert = state["raw_telemetry_alert"]
    # Logic to query SCADA context and format symptoms
    symptoms = [f"Vibration spike {alert['vibration_hz']}Hz at {alert['rpm']}
    RPM"]
    return {"semantic_symptoms": symptoms}

def diagnostic_node(state: MaintenanceState) -> MaintenanceState:
    """Uses RAG to determine root cause from symptoms."""
    symptoms = state["semantic_symptoms"]
    # Tool call: query_vector_db(symptoms, asset_id)
    # LLM reasoning to synthesize diagnosis
    diagnosis = "Inner race bearing fault detected."
    parts = ["BRG-7742-GEN"]
    return {"diagnosis": diagnosis, "required_parts": parts, "confidence_score
    ": 0.92}

def planning_node(state: MaintenanceState) -> MaintenanceState:
    """Checks ERP for parts and MES for scheduling."""
    parts = state["required_parts"]
    # Tool call: check_erp_inventory(parts)
    # Tool call: get_mes_schedule(state["asset_id"])
    # LLM reasoning to find optimal window
    schedule = "Maintenance scheduled for Shift 3 changeover (02:00 AM)."
    return {"parts_available": True, "proposed_schedule": schedule}

def execution_node(state: MaintenanceState) -> MaintenanceState:
    """Generates work orders and adjusts machine parameters."""
    if not state.get("hitl_approved", False):
        return state # Wait for human approval

    # Tool call: create_eam_work_order(...)
    # Tool call: adjust_scada_parameter(asset_id, "feed_rate", 0.85)
    plan = {"status": "Work Order #9942 Created", "parameter_adjusted": True}
    return {"execution_plan": plan}

# 4. Build the Graph Edges
workflow.add_node("Perception", perception_node)
workflow.add_node("Diagnostic", diagnostic_node)
workflow.add_node("Planning", planning_node)
workflow.add_node("Execution", execution_node)

workflow.set_entry_point("Perception")
workflow.add_edge("Perception", "Diagnostic")
workflow.add_edge("Diagnostic", "Planning")
workflow.add_edge("Planning", "Execution")
workflow.add_edge("Execution", END)

```

```
# Compile the graph
app = workflow.compile()
```

This graph structure ensures deterministic execution flow while allowing the LLM within each node to handle the non-deterministic reasoning required for complex diagnostics and scheduling. The separation of concerns prevents a single prompt from becoming overwhelmed by the sheer volume of tools and context required for end-to-end maintenance orchestration.

1.3 Multi-Modal Telemetry Ingestion and Contextual Grounding

One of the most significant challenges in architecting Agentic AI for manufacturing is the impedance mismatch between industrial telemetry and Large Language Models. LLMs are fundamentally semantic engines; they process text and tokens. Industrial assets, however, generate high-frequency, multi-modal, non-semantic data.

A modern CNC machine or industrial pump might generate 10kHz vibration data (10,000 data points per second), continuous thermal imaging streams, and acoustic emission waveforms. If we attempt to feed raw numerical arrays of vibration data directly into an LLM's context window, the model will fail catastrophically. It will consume massive amounts of tokens, hallucinate patterns, and fail to extract meaningful engineering insights. Furthermore, the latency and cost of sending gigabytes of raw time-series data to a cloud-based LLM API are prohibitive.

1.3.1 The Semantic Translation Layer

To solve this, we must implement a *Semantic Translation Layer* at the edge. This layer utilizes traditional, lightweight Machine Learning models (such as 1D Convolutional Neural Networks for vibration, or Autoencoders for anomaly detection) deployed directly on the IoT gateway or edge server. These edge models process the high-frequency data locally and output low-frequency, semantic summaries.

Instead of sending an array of 100,000 floats representing a vibration waveform, the edge model performs a Fast Fourier Transform (FFT), identifies the dominant frequencies, and sends a semantic payload to the Perception Agent:

Listing 1.3: Semantic JSON payload generated by the Edge Translation Layer

```
{
  "asset_id": "PUMP_CENTRIFUGAL_04",
  "timestamp": "2026-06-21T18:10:00Z",
  "event_type": "ANOMALY_DETECTED",
  "semantic_features": [
    "Harmonic peak detected at 2X running speed (3600 RPM).",
    "Overall vibration velocity increased by 4.2 mm/s over 1 hour.",
    "Bearing housing temperature is 15C above historical baseline.",
    "Acoustic emission indicates high-frequency cavitation noise."
  ],
  "current_state": {
    "fluid_viscosity": "high",
    "discharge_pressure": "120 PSI"
  }
```

```
}
```

This semantic payload is highly token-efficient and perfectly suited for LLM reasoning. The LLM can immediately correlate "2X running speed harmonic" with "misalignment" based on its pre-training and RAG-augmented knowledge base.

1.3.2 Context Engineering and RAG for Maintenance

Once the telemetry is translated into semantic text, the Diagnostic Agent must ground its reasoning in the specific reality of the factory. A generic LLM might know that pumps cavitate, but it does not know the specific maintenance history of `PUMP_CENTRIFUGAL_04`.

We achieve this grounding through Context Engineering and Retrieval-Augmented Generation (RAG). We construct a specialized Vector Database (e.g., using a cloud-based Vector Search) that ingests and embeds:

- **OEM Manuals:** PDF manuals detailing troubleshooting steps, exploded view diagrams (processed via multimodal embeddings), and torque specifications.
- **Historical Work Orders:** Past maintenance records from the ERP system. This is crucial for capturing tribal knowledge. If a technician previously noted, *"Replaced seal, but vibration persisted until we shimmed the baseplate,"* the agent needs access to this unstructured text.
- **FMEA Tables:** Structured Failure Mode and Effects Analysis data detailing the probability and severity of specific failure modes.

When the Diagnostic Agent receives the semantic payload, it formulates a search query: *"Centrifugal pump 2X running speed harmonic vibration with elevated temperature and cavitation noise."* The Vector DB returns the most relevant historical cases and manual excerpts. The agent synthesizes this context to produce a highly accurate, asset-specific diagnosis. This fusion of real-time semantic telemetry with historical RAG context is what elevates the system from a simple anomaly detector to an expert diagnostic reasoning engine.

1.4 The Agentic Twin: Counterfactual Simulation and Safe Execution

The most powerful—and most dangerous—capability of an Agentic Predictive Maintenance system is its ability to execute physical changes in the real world. If an agent determines that a machine will fail before the end of the shift, it may propose adjusting the machine's operating parameters (e.g., reducing spindle speed, lowering feed rates, or decreasing pressure) to extend the Remaining Useful Life (RUL) and save the production run.

However, industrial environments are highly constrained. An agent cannot blindly alter a recipe parameter without understanding the cascading effects on product quality, cycle time, and downstream bottlenecks. To solve this, we implement the **Agentic Twin** pattern.

1.4.1 Counterfactual Simulation

An Agentic Twin is an active digital twin that serves as a simulation sandbox for the AI agent. Before the Execution Agent sends any command to the SCADA system, it must first propose the action to the Agentic Twin and request a counterfactual simulation.

Consider a scenario where a CNC machine is overheating. The agent proposes: `adjust_parameter(feed_rate_multiplier=0.85)`. The Agentic Twin runs a deterministic physics and scheduling simulation and returns the following impact analysis to the agent:

- **Asset Health Impact:** Thermal load reduced by 18%. RUL extended by 4.5 hours (sufficient to reach the end of the shift).
- **Production Impact:** Cycle time increased by 12 seconds per part. Shift throughput reduced by 40 units.
- **Constraint Check:** The reduced throughput of 460 units still exceeds the daily minimum quota of 450 units. **PASS.**

If the simulation reveals that the proposed action would violate a critical constraint (e.g., causing a downstream assembly line to starve for parts), the agent receives a failure response and must loop back to the Planning phase to generate an alternative strategy (e.g., proposing an immediate emergency maintenance window instead).

1.4.2 Safety Guardrails and the HITL Gateway

Even with rigorous simulation, autonomous actuation in heavy industry requires strict safety guardrails. These guardrails are implemented outside the LLM, at the API gateway level, ensuring that even if the model hallucinates, catastrophic physical actions are impossible.

1. **Bounded Action Spaces:** The tools provided to the agent have hard-coded limits. The `adjust_feed_rate` tool might enforce a rule: `0.70 ⩽ multiplier ⩽ 1.0`. If the agent attempts to set the multiplier to 2.0, the API rejects the call immediately. 2. **State-Based Lockouts:** Certain actions are only permitted in specific machine states. An agent cannot initiate a tool-change sequence if the spindle is currently engaged in cutting material. 3. **Human-in-the-Loop (HITL) Gateway:** For high-stakes interventions, the architecture mandates human approval. The agent acts as an expert advisor, compiling a "Maintenance Briefing" that includes the semantic telemetry, the RAG-backed diagnosis, the proposed action, and the Agentic Twin simulation results.

Listing 1.4: Implementing a HITL Gateway for High-Stakes Actuation

```
def hitl_approval_gateway(proposed_action: dict, simulation_results: dict) ->
    bool:
    """
    Routes the agent's proposed action to a human engineer for approval.
    """
    # 1. Check if action requires HITL based on risk matrix
    if proposed_action["risk_level"] < THRESHOLD_LOW_RISK:
        return True # Auto-approve low risk actions (e.g., logging a ticket)

    # 2. Format the briefing for the human engineer
    briefing = f"""
```

```
    URGENT MAINTENANCE APPROVAL REQUIRED
    Asset: {proposed_action['asset_id']}
    Diagnosis: {proposed_action['diagnosis']}
    Proposed Action: {proposed_action['action_description']}
    Simulated Impact: {simulation_results['summary']}
    """

    # 3. Dispatch to human interface (e.g., messaging platforms or a custom
    MES dashboard)
    approval_status = dispatch_to_human_ui(briefing)

    # 4. Block execution until human clicks 'Approve' or 'Reject'
    return wait_for_human_response(approval_status)
```

This HITL pattern ensures compliance with industrial safety regulations (such as OSHA standards and the EU AI Act's requirements for high-risk AI systems) while still capturing 90% of the efficiency gains of autonomy. The human engineer is no longer spending hours digging through data to find the problem; they are simply reviewing and approving a fully synthesized, simulated, and optimized plan presented by the agentic system.

Key Takeaways

- Agentic Predictive Maintenance shifts the paradigm from merely predicting failures to autonomously reasoning about, planning, and executing maintenance strategies within production constraints.
- A Multi-Agent System (MAS) architecture separates concerns into Perception, Diagnostic, Planning, and Execution roles, preventing context overload and ensuring deterministic workflows.
- High-frequency, multi-modal telemetry must be processed by edge ML models into a "Semantic Translation Layer" before being fed into an LLM's context window.
- The "Agentic Twin" pattern allows agents to run counterfactual simulations, ensuring proposed parameter adjustments optimize OEE without violating production quotas.
- Autonomous physical actuation requires strict, API-level bounded action spaces and Human-in-the-Loop (HITL) approval gateways to guarantee industrial safety.

Learning Objectives

- Understand the transition from static Advanced Planning and Scheduling (APS) systems to dynamic, agent-driven Theory of Constraints (ToC) optimization.
- Architect a real-time Semantic Factory Graph that allows Large Language Models to reason over complex, multi-stage manufacturing topologies.
- Implement a Multi-Agent System (MAS) where Workstation Agents and Buffer Agents negotiate routing and capacity in real-time to mitigate wandering bottlenecks.
- Integrate the Agentic Twin (introduced in Chapter 260) to run counterfac-

tual discrete-event simulations before actuating physical routing changes via SCADA/MES.
- Design reward functions and context-window management strategies for high-frequency industrial orchestration.

1.5 The Agentic Theory of Constraints (ToC)

In the previous chapter, we explored how Agentic Predictive Maintenance utilizes the Semantic Translation Layer and the Agentic Twin to foresee equipment failures and safely adjust machine parameters to extend Remaining Useful Life (RUL). However, keeping a single machine running is only a localized optimization. In complex manufacturing environments—particularly High-Mix, Low-Volume (HMLV) facilities, semiconductor fabs, and automotive assembly lines—the ultimate metric is global throughput. Optimizing a non-bottleneck machine is, as Dr. Eliyahu M. Goldratt famously articulated in his Theory of Constraints (ToC), an illusion of efficiency.

Traditional manufacturing relies on Advanced Planning and Scheduling (APS) systems and Manufacturing Execution Systems (MES) to manage throughput. These systems operate on deterministic, batch-processed algorithms (such as Mixed-Integer Linear Programming). They assume a static factory state: they calculate the critical path, identify the primary bottleneck, and generate a rigid production schedule for the shift.

The reality of the factory floor, however, is highly stochastic. A minor tool wear issue on a CNC lathe, a 45-second delay in an Automated Guided Vehicle (AGV) delivery, or a slight variation in raw material hardness can cause the bottleneck to shift dynamically. This phenomenon is known as the *wandering bottleneck*. When a bottleneck wanders, the static APS schedule becomes obsolete within minutes of the shift starting. Work-in-Progress (WIP) accumulates in front of the new, unpredicted constraint (blocking upstream machines), while downstream machines starve for parts.

Figure 1.3: The Wandering Bottleneck: Traditional APS systems assume a static constraint, whereas Agentic systems continuously track and respond to dynamic shifts in line constraints caused by micro-stoppages.

The **Agentic Throughput & Bottleneck Planner** represents a paradigm shift from static scheduling to dynamic, continuous orchestration. By deploying Large Language Models (LLMs) equipped with graph-reasoning capabilities, we can create an intelligent orchestrator that continuously ingests the state of the factory, identifies emerging constraints in real-time, and autonomously negotiates routing, buffer allocation, and speed adjustments to maintain optimal flow.

This approach applies Little's Law ($L = \lambda W$, where L is WIP, λ is throughput, and W is lead time) dynamically. Instead of trying to force the factory to adhere to a brittle schedule, the agentic system constantly adjusts the flow (λ) and the buffers (L) to minimize lead time (W) in response to the actual, observed state of the physical plant. The LLM's ability to process complex, unstructured rules (e.g., "Product A cannot be routed to Machine 4 if Machine 4 hasn't been calibrated in the last 12 hours") alongside structured telemetry makes it vastly superior to traditional heuristic-based routing engines.

However, LLMs cannot natively read programmable logic controller (PLC) registers or raw OPC UA data streams. To enable agentic reasoning over a physical factory, we must first construct a representation of the factory that the LLM can understand: the Semantic Factory Graph.

1.6 Architecting the Factory Graph and State Representation

The foundational architecture for the Throughput Planner relies on translating the physical topology of the manufacturing line into a dynamic, token-efficient semantic graph. If we attempt to feed raw JSON dumps from an MES into an LLM's context window, we will rapidly exhaust the token limit, introduce unacceptable latency, and degrade the model's reasoning performance due to the "lost in the middle" phenomenon.

Instead, we utilize the *Semantic Translation Layer* (discussed in Chapter 260) to aggregate high-frequency telemetry into low-frequency state changes, and we map these states onto a directed graph. In this graph:

- **Nodes** represent Workstations (machines, manual assembly benches) and Buffers (storage areas, accumulation conveyors).
- **Edges** represent the physical material flow paths (conveyors, AGV routes, gantry cranes) and their current capacities.
- **State Properties** include current status (running, down, blocked, starving), WIP count, cycle time variance, and active product mix.

Figure 1.4: The Agentic Ingestion Pipeline: Raw OPC UA telemetry is translated into a semantic graph state, which is injected into the Global Orchestrator Agent's context window for reasoning.

To implement this, we use a state management framework like LangGraph combined with Pydantic for strict schema validation. The factory state is updated via an event-driven architecture (e.g., MQTT messages from the edge gateway triggering state mutations).

The following code demonstrates how to define this semantic graph state in Python. Notice how the state is designed to be highly readable for an LLM, focusing on the semantic meaning of the data rather than raw sensor values.

Listing 1.5: Defining the Semantic Factory Graph State using Pydantic and LangGraph

```
from typing import List, Dict, Literal, Optional
from pydantic import BaseModel, Field
from langgraph.graph import StateGraph, END

# 1. Define the Semantic Schema for Nodes and Edges
class WorkstationState(BaseModel):
    node_id: str = Field(description="Unique identifier for the workstation")
    status: Literal["RUNNING", "DOWN", "BLOCKED", "STARVING", "CHANGEOVER"]
```

```python
    current_wip: int = Field(description="Number of units currently at this station")
    avg_cycle_time_sec: float = Field(description="Moving average cycle time")
    health_score: float = Field(description="0.0 to 1.0, provided by Predictive Maintenance Agent")

class BufferState(BaseModel):
    buffer_id: str
    capacity: int
    current_occupancy: int
    utilization_pct: float

class FactoryGraphState(BaseModel):
    timestamp: str
    workstations: Dict[str, WorkstationState]
    buffers: Dict[str, BufferState]
    active_bottleneck_id: Optional[str] = Field(default=None, description="Currently identified constraint")
    global_throughput_uph: float = Field(description="Units per hour")

# 2. Define the Agentic State Reducer (LangGraph)
def update_factory_state(state: FactoryGraphState, new_telemetry: dict) -> FactoryGraphState:
    """
    Reducer function that takes incoming MQTT telemetry and updates the semantic graph.
    This runs continuously outside the LLM context to maintain the 'digital twin' state.
    """
    # Example: Update workstation status based on incoming event
    ws_id = new_telemetry.get("node_id")
    if ws_id in state.workstations:
        ws = state.workstations[ws_id]
        ws.status = new_telemetry.get("status", ws.status)
        ws.current_wip = new_telemetry.get("wip", ws.current_wip)

        # Simple heuristic to flag potential bottlenecks before LLM reasoning
        if ws.status == "BLOCKED" and ws.current_wip > 10:
            state.active_bottleneck_id = ws_id

    return state

# Initialize the LangGraph State
workflow = StateGraph(FactoryGraphState)
# (Node and Edge definitions for the reasoning loop will follow in Section 3)
```

By maintaining this state object, we decouple the high-frequency physical world from the low-frequency cognitive world. The LLM does not need to process a stream of 10Hz sensor data. Instead, it is invoked only when the `update_factory_state` function detects a significant semantic shift—for example, when a workstation transitions from `RUNNING` to `BLOCKED`, or when a buffer's `utilization_pct` exceeds 90%.

When invoked, the LLM receives a serialized, token-efficient JSON representation of `FactoryGraphState`. Because models like Gemini 3.5 Pro possess deep spatial and logical reasoning capabilities, they can "read" this JSON graph, understand the topology of the factory, and immediately identify that if Node C is blocked and Node D is starving, the constraint must lie on the edge between them.

1.7 Multi-Agent Negotiation for Dynamic Line Balancing

While a single Global Orchestrator Agent can manage a small production line, scaling this architecture to a massive facility (e.g., an automotive plant with hundreds of interconnected cells) requires a Multi-Agent System (MAS). A monolithic agent would suffer from context bloat and excessive latency.

Instead, we deploy a hierarchical MAS architecture utilizing the Actor Model pattern. In this design, we instantiate:

- **Workstation Agents:** Localized agents responsible for a specific machine or cell. Their goal is to maximize their own local OEE (Overall Equipment Effectiveness). They monitor local telemetry and manage local buffers.
- **Transport Agents:** Agents controlling AGV fleets or conveyor routing logic.
- **Global Orchestrator Agent:** The meta-agent that oversees the entire graph. Its sole objective is to maximize global throughput (λ). It resolves conflicts between Workstation Agents.

This architecture shines during stochastic disruptions. Consider a scenario where a primary CNC milling machine (`CNC_A`) experiences accelerated tool wear. The Predictive Maintenance Agent (from Chapter 260) intervenes and slows the feed rate by 15% to prevent a catastrophic tool break.

Immediately, `CNC_A` becomes a wandering bottleneck. Its cycle time increases. The Workstation Agent for `CNC_A` detects that its inbound buffer is filling up rapidly and broadcasts a distress signal to the MAS network.

Figure 1.5: Multi-Agent Negotiation: Local Workstation Agents broadcast their state (starving/blocked), allowing the Global Orchestrator to dynamically reroute material flow to balance the line.

Simultaneously, an older, less efficient machine (`CNC_B`) is currently idle. The Workstation Agent for `CNC_B` is broadcasting a "Starving" status. In a traditional APS system, rerouting WIP from A to B mid-shift would require a manual planner to intervene, update the ERP, and issue new paper travelers—a process taking hours.

In the Agentic Planner, the agents negotiate in real-time. The Global Orchestrator receives the signals, evaluates the graph, and initiates a routing change.

Listing 1.6: Prompt Template for the Global Orchestrator Agent resolving a routing conflict

```
orchestrator_prompt = """
```

```
You are the Global Throughput Orchestrator for Factory Line Alpha.
Your objective is to maximize global units per hour (UPH).

CURRENT FACTORY STATE:
{factory_graph_json}

ACTIVE ALERTS:
- Workstation CNC_A reports: "Inbound buffer at 95%. Cycle time degraded by 15% due to tool wear mitigation."
- Workstation CNC_B reports: "Status STARVING. Inbound buffer at 0%."

AVAILABLE ACTIONS:
1. route_wip(source_buffer, destination_node, quantity)
2. adjust_conveyor_speed(conveyor_id, speed_pct)
3. request_maintenance(node_id, priority)

CONSTRAINTS:
- CNC_B has a 5% higher defect rate than CNC_A for this specific part number.
- Rerouting via AGV takes 4 minutes.

Analyze the state and output a JSON execution plan using the available actions to resolve the bottleneck at CNC_A without starving downstream assembly.
"""
```

The LLM processes this prompt and reasons through the trade-offs. It calculates that the 4-minute AGV transit time and the 5% higher defect rate at `CNC_B` are mathematically preferable to allowing the entire downstream assembly line to starve due to `CNC_A`'s degraded throughput. The Orchestrator outputs a structured JSON tool call invoking `route_wip("BUFFER_A_IN", "CNC_B", 20)`.

This multi-agent negotiation happens in seconds, continuously rebalancing the line as micro-stoppages occur, effectively smoothing the flow of production and maximizing the utilization of available capacity.

1.8 Counterfactual Simulation and Actuation (The Agentic Twin in Action)

The decision to reroute WIP or adjust line speeds carries significant physical and financial risk. If the Global Orchestrator hallucinates a routing command that sends heavy castings to a conveyor rated only for lightweight plastics, the result is catastrophic equipment damage. Furthermore, complex routing changes can cause unforeseen cascading effects, such as creating a worse bottleneck further downstream.

To mitigate this, the Throughput Planner must never actuate directly on the physical plant based solely on its initial reasoning. It must utilize the **Agentic Twin** pattern. Before any tool call (like `route_wip`) is sent to the MES or SCADA system, it is intercepted by the orchestration layer and routed to a Discrete-Event Simulation (DES) engine.

1.8.1 Integrating Discrete-Event Simulation (DES)

While LLMs are excellent at semantic reasoning and heuristic problem-solving, they are notoriously poor at complex, multi-step deterministic mathematics (like calculating the exact queue accumulation over 4 hours across 50 interconnected nodes). DES engines (such as Siemens Plant Simulation, AnyLogic, or open-source Python libraries like SimPy) are purpose-built for this exact mathematical task.

The Agentic Planner bridges this gap by using the LLM to generate *hypotheses* and the DES engine to *validate* them.

When the Orchestrator proposes `route_wip("BUFFER_A_IN", "CNC_B", 20)`, the system translates this command into a simulation scenario. The Agentic Twin spins up a fast-forward simulation of the next 4 hours of production, applying the proposed routing change.

The DES engine returns a deterministic impact report to the LLM:

- **Baseline (Do Nothing):** Projected Shift Throughput = 410 units. Downstream starvation begins in 22 minutes.
- **Proposed Action (Reroute to CNC_B):** Projected Shift Throughput = 445 units. Downstream starvation averted. However, Buffer_C_OUT will overflow in 110 minutes due to CNC_B's slower processing speed.

Figure 1.6: Cumulative Throughput Recovery: The Agentic Planner leverages counterfactual simulation to dynamically reroute production around micro-stoppages, recovering throughput that traditional static APS systems permanently lose.

1.8.2 The Reflection and Replanning Loop

Upon receiving the simulation results, the Orchestrator Agent enters a *Reflection Loop*. It recognizes that its initial plan solved the immediate bottleneck but created a secondary constraint (Buffer_C_OUT overflow).

The LLM is prompted to refine its plan: *"Your previous action caused a simulated overflow at Buffer_C_OUT at T+110 mins. Propose an updated execution plan to mitigate this secondary constraint."*

The agent reasons that to prevent the overflow, it must increase the pull rate from Buffer_C_OUT. It issues a revised plan: 1. `route_wip("BUFFER_A_IN", "CNC_B", 20)` 2. `adjust_conveyor_speed("CONV_C_TO_D", 1.15)` (Increase downstream takeaway speed by 15%).

The Agentic Twin simulates this revised plan. The result: Projected Throughput = 452 units, no buffer overflows, all constraints satisfied.

Only after the simulation returns a clean, constraint-free validation does the system pass the execution plan to the Human-in-the-Loop (HITL) Gateway. The human Production Supervisor receives a concise briefing on their tablet: *"Agent proposes rerouting 20 units to CNC_B and increasing Conveyor C speed to mitigate tool wear on CNC_A. Simulated impact: +42 units for the shift. Approve?"*

Once approved, the API gateway translates the semantic tool calls into the specific OPC UA write commands or REST API calls required by the MES to actuate the physical changes on the floor.

This integration of LLM heuristic reasoning with DES deterministic validation represents the state-of-the-art in industrial AI. It provides the flexibility and adaptability of human-like reasoning, bounded by the rigorous mathematical safety required in heavy manufacturing. The result is a highly resilient production line capable of autonomously navigating the chaos of stochastic disruptions to maximize global throughput.

Key Takeaways

- **Wandering Bottlenecks:** Traditional APS systems fail because they assume static constraints. Agentic systems continuously monitor the factory graph to identify and mitigate dynamic, shifting bottlenecks.
- **Semantic Factory Graph:** LLMs cannot process raw high-frequency telemetry. The physical plant must be abstracted into a token-efficient graph state (Nodes, Edges, WIP levels) for effective reasoning.
- **Multi-Agent Actor Model:** Scaling to large factories requires decentralization. Workstation Agents manage local OEE and negotiate with a Global Orchestrator Agent that optimizes for global throughput.
- **LLM + DES Synergy:** LLMs generate heuristic routing hypotheses; Discrete-Event Simulation (DES) engines validate them mathematically. This prevents catastrophic cascading failures.
- **Safe Actuation:** Physical changes to the line must pass through the Agentic Twin simulation and a Human-in-the-Loop (HITL) gateway before being executed via MES/SCADA APIs.

Learning Objectives

- Understand the limitations of static manufacturing recipes and traditional Design of Experiments (DoE) in highly variable production environments.
- Architect an Agentic Parameter Optimizer that combines the semantic reasoning of Large Language Models (LLMs) with the mathematical rigor of Bayesian Optimization.
- Implement strict Policy-as-Code guardrails and bounded action spaces to ensure physical safety and prevent catastrophic machine failure.
- Design a Human-in-the-Loop (HITL) gateway that translates agentic recommendations into actionable, safe PLC write commands via OPC UA.
- Evaluate the ROI and latency trade-offs of deploying cognitive optimization loops at the industrial edge versus the cloud.

1.9 The Shift from Static Recipes to Dynamic Agentic Control

In the context of industrial manufacturing, a "recipe" is the specific combination of machine parameters, setpoints, and timings required to transform raw materials into a finished product. Whether it is the temperature profile of a semiconductor diffusion furnace, the injection speed and holding pressure of a plastic injection molding machine, or the chemical mix ratios in a continuous process reactor, the recipe dictates the quality, yield, and energy efficiency of the operation. Historically, these recipes have been static, derived through painstaking manual processes, and rigidly enforced by Programmable Logic Controllers (PLCs). The transition from these static paradigms to dynamic, agentic control represents one of the most profound leaps in Industry 4.0, moving from systems that merely execute instructions to systems that autonomously seek optimal states.

The traditional approach to recipe generation relies heavily on Design of Experiments (DoE) and Taguchi methods. Process engineers run dozens or hundreds of physical trials, systematically varying parameters (e.g., temperature, pressure, speed) to map the response surface and identify the optimal operating window. Once this "golden recipe" is found, it is hardcoded into the Manufacturing Execution System (MES) or directly into the machine's PLC. The underlying assumption of a static recipe is that the manufacturing environment is perfectly controlled and that raw materials are perfectly consistent. In reality, the physical world is stochastic. Ambient humidity fluctuates between morning and afternoon shifts; the viscosity of a polymer batch varies slightly from the previous lot; machine components wear down, altering their thermal transfer coefficients.

When these unmodeled variances occur, the static recipe fails. The machine continues to execute the exact same parameters, but because the inputs or environmental conditions have drifted, the output quality degrades, leading to increased scrap rates, rework, and wasted energy. The traditional response to this drift is reactive: an operator notices a defect, flags a process engineer, and the engineer manually tweaks the parameters based on intuition and experience—a process that is slow, unscalable, and highly dependent on tribal knowledge.

Advanced Process Control (APC) and Model Predictive Control (MPC) attempted to solve this by using mathematical models to predict future states and adjust parameters dynamically. However, MPC requires highly accurate, explicit mathematical models of the physical process, which are notoriously difficult and expensive to build and maintain for complex, non-linear manufacturing processes. Furthermore, traditional APC systems lack semantic understanding; they cannot read an operator's shift log that says, "The parts are coming out with slight flash on the left edge, and the machine sounds louder than usual," and translate that unstructured observation into a targeted parameter search.

Figure 1.7: Static vs. Dynamic Agentic Recipe Control. While static recipes fail under environmental drift, Agentic Optimizers continuously adjust setpoints by combining real-time telemetry with semantic reasoning and mathematical optimization.

This is where the Agentic Parameter Optimizer introduces a paradigm shift. By leveraging Large Language Models (LLMs) as cognitive orchestrators, we can bridge the gap between unstructured human observations, high-frequency machine telemetry, and rigorous mathematical optimization. The LLM does not replace the PID controllers that manage the micro-second physical actuation; rather, it acts as a Supervisory Control agent. It sits above the PLC layer, continuously monitoring the state of the process.

The true power of the agentic approach lies in its ability to orchestrate specialized tools. An LLM is inherently poor at calculating the exact gradient descent required to optimize a 15-dimensional parameter space. However, an LLM is exceptionally good at understanding the semantics of the manufacturing process, formulating the optimization problem, and writing the code to invoke a specialized mathematical solver (such as a Bayesian Optimization engine).

For example, if a vision inspection system detects "short shots" (incomplete filling of a mold) in an injection molding process, the Agentic Optimizer receives this semantic flag. Drawing on its grounded knowledge base of polymer physics and historical machine data, the agent reasons: "A short shot indicates insufficient material volume or premature freezing. The potential corrective actions are increasing injection pressure, increasing melt temperature, or increasing shot size. I will invoke the Bayesian Optimization tool, constraining the search space for Melt Temperature to [230°C, 245°C] and Injection Pressure to [1200 bar, 1400 bar], to find the optimal new setpoints that minimize the short shot probability without exceeding the machine's safety limits."

This capability transforms the factory floor. Instead of relying on a scarce team of process engineers to manually tune machines across hundreds of production lines, the enterprise deploys a fleet of Agentic Optimizers. These agents continuously run micro-experiments, safely exploring the parameter space within bounded limits, learning the unique idiosyncrasies of each specific machine, and dynamically adapting the recipe to compensate for raw material variance and environmental drift. The result is a self-healing manufacturing process that maximizes yield, minimizes scrap, and operates at peak efficiency regardless of external stochasticity.

1.10 Architecting the Parameter Optimization Agent

Building an Agentic Parameter Optimizer requires a robust architecture that seamlessly integrates the cognitive reasoning capabilities of an LLM with the deterministic, high-frequency reality of industrial control systems. This architecture must bridge the IT (Information Technology) and OT (Operational Technology) divide, translating semantic goals into precise, machine-readable commands. The system is typically structured across three distinct layers: the Physical Layer (OT), the Integration Layer (Data & Simulation), and the Cognitive Layer (Agentic Orchestration).

At the foundation is the **Physical Layer**, comprising the PLCs, sensors, actuators, and the physical machinery itself. This layer operates in real-time (often sub-millisecond latency) and is strictly deterministic. It executes the setpoints provided to it. Above this sits the **Integration Layer**, which acts as the translation and buffering zone. This layer utilizes industrial communication protocols, predominantly OPC UA (Open Platform Communications Unified Architecture), to securely expose machine tags (data points) to higher-level systems. It also includes Time-Series Historians (like OSIsoft PI or InfluxDB) that store high-frequency telemetry, and Discrete Event Simulators or Digital Twins that model the physical physics of the process.

The **Cognitive Layer** is where the Agentic Optimizer resides. This layer operates at a lower frequency (e.g., evaluating the process every few minutes or upon receiving a defect alert) but possesses high semantic understanding. We implement this layer using stateful graph frameworks like LangGraph or CrewAI, allowing the agent to maintain memory of past interventions, reason over current telemetry, and execute multi-step optimization plans.

Figure 1.8: Agentic Parameter Optimizer Dataflow. The architecture separates the high-frequency deterministic control (Physical Layer) from the low-frequency semantic reasoning (Cognitive Layer), bridged by secure OPC UA integration and safety gateways.

The core of the Cognitive Layer is the LLM Orchestrator, which is equipped with specialized tools. As established in previous chapters, LLMs should not perform complex mathematics directly. Instead, the agent is provided with a `BayesianOptimizationTool`. When the agent detects a process drift, it formulates the boundaries of the problem and passes them to the tool. Bayesian Optimization is particularly suited for manufacturing because it is highly sample-efficient; it builds a probabilistic surrogate model (usually a Gaussian Process) of the unknown objective function and uses an acquisition function to decide where to sample next, minimizing the number of physical trials required to find the optimum.

Let us examine the LangGraph implementation of this reasoning loop. The state of the graph must capture the current machine telemetry, the identified defect or drift, the proposed parameter search space, and the results of the mathematical optimization.

Listing 1.7: LangGraph State and Node Definition for the Agentic Parameter Optimizer

```
from typing import Dict, List, Optional, TypedDict
from pydantic import BaseModel, Field
from langgraph.graph import StateGraph, END

# 1. Define the State Schema
class OptimizationState(TypedDict):
    machine_id: str
    current_telemetry: Dict[str, float]
    detected_issue: str
    search_space: Optional[Dict[str, tuple]] # e.g., {"temp": (200, 250)}
    proposed_parameters: Optional[Dict[str, float]]
    simulation_result: Optional[str]
    approval_status: str # "PENDING", "APPROVED", "REJECTED"

# 2. Define the Reasoning Node (LLM formulates the problem)
def formulate_search_space(state: OptimizationState) -> OptimizationState:
    """LLM analyzes the issue and defines the bounds for the math solver."""
    prompt = f"""
    Machine: {state['machine_id']}
    Issue: {state['detected_issue']}
    Current Telemetry: {state['current_telemetry']}

    Based on polymer physics, identify which parameters to adjust to fix the
    issue.
    Output a JSON dictionary defining the search space bounds (min, max) for
    the Bayesian Optimizer.
    Do not exceed absolute machine safety limits.
    """
    # In production, invoke LLM with structured output (JSON)
    # Mock LLM response for illustration:
    llm_response = {"melt_temp": (230.0, 245.0), "injection_speed": (45.0,
    60.0)}
    state['search_space'] = llm_response
    return state

# 3. Define the Math Tool Node (Executes Bayesian Optimization)
def run_bayesian_optimizer(state: OptimizationState) -> OptimizationState:
    """Invokes external Python ML library (e.g., BoTorch/Ax) to find optimal
    points."""
```

```
    bounds = state['search_space']
    # ... (Code to initialize Gaussian Process and run acquisition function)
    ...
    # Mock result from the math solver:
    optimal_params = {"melt_temp": 238.5, "injection_speed": 52.0}
    state['proposed_parameters'] = optimal_params
    return state

# 4. Build the Graph
workflow = StateGraph(OptimizationState)
workflow.add_node("formulate_space", formulate_search_space)
workflow.add_node("run_optimizer", run_bayesian_optimizer)
# (Simulation and HITL nodes would follow)

workflow.add_edge("formulate_space", "run_optimizer")
# ...
```

In this architecture, the LLM acts as the domain expert. It knows that to fix a "short shot," it should look at temperature and speed, not cooling time. It sets the intelligent boundaries `(230.0, 245.0)`. The Bayesian Optimizer then takes over, doing the heavy mathematical lifting to find the exact optimal point `238.5` within that space based on historical data correlations.

This separation of concerns is critical. It prevents the LLM from hallucinating arbitrary numbers while leveraging its vast semantic knowledge base to guide the mathematical solver efficiently. Furthermore, by maintaining this process within a stateful graph, the agent can iterate. If the Bayesian Optimizer fails to find a solution that satisfies the quality constraints, the graph can loop back to the `formulate_search_space` node, prompting the LLM to expand the search space or consider adjusting a different set of parameters (e.g., holding pressure). This recursive, self-correcting loop mirrors the troubleshooting process of a senior human engineer, but executes in seconds rather than hours.

1.11 Bounded Action Spaces and Safety Guardrails

The introduction of autonomous agents into physical manufacturing environments introduces significant risk. Unlike software environments where a hallucinated output might result in a broken webpage or a failed API call, a hallucinated parameter in a chemical reactor or a CNC machine can result in catastrophic equipment damage, fires, or severe injury to human operators. Therefore, the Agentic Parameter Optimizer cannot be deployed with direct, unfettered write access to the PLC. It must be constrained by rigorous, deterministic safety architectures.

The core principle for safe industrial AI is the implementation of **Bounded Action Spaces** and **Policy-as-Code Guardrails**. The LLM is inherently non-deterministic; we must surround it with deterministic filters. We conceptualize this as a "Safety Funnel" through which every agent-proposed parameter change must pass before it is allowed to actuate physical hardware.

Figure 1.9: The Safety Guardrail Funnel. Non-deterministic LLM proposals are forced through deterministic boundary checks, physics simulations, and human approval gates before reaching the physical machinery.

The first layer of the funnel is the **Hard Constraint Boundary Check**. This is a purely deterministic software layer, often implemented using data validation libraries like Pydantic in Python. This layer codifies the absolute physical limits of the machine, regardless of what the LLM or the Bayesian Optimizer suggests. If a machine's heating element is physically rated for a maximum of 300°C, the boundary check is hardcoded to reject any proposal exceeding 290°C. This layer operates entirely independently of the AI models.

Listing 1.8: Implementing Deterministic Hard Constraints using Pydantic

```python
from pydantic import BaseModel, Field, validator

class InjectionMoldingRecipe(BaseModel):
    melt_temperature_c: float = Field(..., description="Temperature of the polymer melt")
    injection_speed_mms: float = Field(..., description="Speed of the injection ram")
    holding_pressure_bar: float = Field(..., description="Pressure applied after injection")

    @validator('melt_temperature_c')
    def check_temp_limits(cls, v):
        # Absolute physical limits of the machine heater bands
        if not (180.0 <= v <= 280.0):
            raise ValueError(f"Melt temperature {v} exceeds absolute machine safety limits [180, 280].")
        return v

    @validator('holding_pressure_bar')
    def check_pressure_limits(cls, v):
        # Prevent catastrophic mold blowout
        if v > 1500.0:
            raise ValueError(f"Holding pressure {v} exceeds maximum safe mold clamp force.")
        return v

def validate_agent_proposal(proposed_json: dict) -> bool:
    try:
        # Force the LLM's output through the deterministic Pydantic model
        safe_recipe = InjectionMoldingRecipe(**proposed_json)
        return True
    except ValueError as e:
        print(f"AGENT SAFETY VIOLATION BLOCKED: {e}")
        return False
```

If the agent's proposal passes the hard boundary checks, it enters the second layer of the funnel: **Physics and Thermodynamic Validation**. While a parameter might be within the absolute limits of the machine, the *combination* of parameters might be dangerous or yield guaranteed scrap. For example, a high melt temperature combined with a very slow injection speed might cause the polymer to degrade and burn inside the barrel. This layer utilizes the Agentic Twin pattern discussed in Chapter 261. The proposed recipe is sent to a Digital Twin or a physics-based simulation engine (e.g., a simplified finite element analysis or a thermodynamic model). The simulation runs a fast-forward prediction of the process. If the simulation predicts a thermal runaway or a pressure spike that exceeds safety thresholds, the proposal is rejected, and the agent is prompted to replan with the simulation error as feedback.

The final layer of the funnel is the **Human-in-the-Loop (HITL) Gateway**. For highly regulated industries (like pharmaceuticals or aerospace) or for processes with high safety risks, autonomous closed-loop control is often prohibited by compliance standards (such as FDA 21 CFR Part 11). In these scenarios, the Agentic Optimizer operates in an "open-loop" or "copilot" mode. The agent completes its reasoning, passes the boundary checks, passes the simulation, and then presents a highly contextualized recommendation to a human operator via a dashboard.

The dashboard does not just show the new numbers; it explains the *why*. It presents the agent's chain of thought: "Detected a 5% increase in ambient humidity leading to part warping. Simulated a 3°C increase in mold temperature and a 2% decrease in holding pressure. Simulation predicts this will return warp tolerances to baseline. Do you approve writing these setpoints to PLC-04?" The human operator reviews the logic, clicks "Approve," and only then does the API gateway execute the OPC UA write command to alter the physical machine state. This ensures that human accountability is maintained while drastically reducing the cognitive load and time required to troubleshoot the process.

1.12 Real-World Implementation: Injection Molding Case Study

To understand the practical impact and operational mechanics of the Agentic Parameter Optimizer, we examine a real-world deployment in a high-volume plastic injection molding facility producing automotive interior components. Injection molding is a notoriously complex process characterized by high dimensionality; a single machine may have over 40 adjustable parameters, and the process is highly sensitive to environmental factors, polymer batch variations, and machine wear.

In this facility, the baseline operation relied on static recipes developed during initial tooling tryouts. The primary quality metric was dimensional stability (preventing warp) and visual finish (preventing sink marks or flash). The facility experienced a recurring issue: during the transition from the cool night shift to the hot, humid afternoon shift, the scrap rate would reliably spike from a baseline of 2% up to 10-12%. The ambient temperature change altered the cooling dynamics of the mold, and the increased humidity affected the hygroscopic polymer pellets.

Traditionally, a process engineer would spend 2-3 hours each afternoon manually tweaking the cooling time, mold temperature, and holding pressure to stabilize the process, resulting in significant lost production time and wasted material during the tuning phase.

The engineering team deployed an Agentic Parameter Optimizer utilizing the Google Agent Development Kit (ADK) and Vertex AI, integrated with the factory's Ignition SCADA system via OPC UA. The architecture was deployed at the industrial edge (using Google Distributed Cloud Edge) to ensure low latency and continuous operation even if the external internet connection was severed, a critical requirement for OT environments.

Figure 1.10: Scrap Rate Reduction. The Agentic Optimizer dynamically adjusts to environmental drift, maintaining a stable, low scrap rate compared to the highly volatile manual tuning approach.

The operational flow of the deployed agent was as follows: 1. **Sense:** The SCADA system continuously streamed machine telemetry (barrel temperatures, injection pressures, cycle times) and environmental data (ambient temp, humidity) to the edge historian. An inline automated optical inspection (AOI) camera measured the dimensions of every produced part. 2. **Trigger:** When the AOI system detected that the part dimensions were drifting toward the upper control limit (indicating impending warp defects), it triggered the Agentic Optimizer. 3. **Reason & Plan:** The LLM (Gemini 3.5 Pro, optimized for edge inference) ingested the recent telemetry and the defect trend. It reasoned that the increased ambient temperature was reducing the effectiveness of the mold cooling circuit. It formulated a search space, restricting the Bayesian Optimizer to adjust only the Mold Temperature Setpoint (within a tight $\pm 5°C$ band) and the Cooling Time (within a ± 2 second band). 4. **Optimize:** The Bayesian Optimization tool ran a rapid surrogate model calculation, determining that dropping the mold temperature by 2.5°C and increasing cooling time by 0.8 seconds would yield the highest probability of returning the dimensions to nominal without significantly impacting overall cycle time. 5. **Validate & Actuate:** The proposed parameters passed the Pydantic hard-constraint checks. Because this specific process was deemed low safety risk, the system was configured for closed-loop control. The agent bypassed the HITL gateway and directly issued an OPC UA write command to the machine's PLC.

The entire cycle—from detecting the drift to actuating the new recipe—took less than 15 seconds. The agent continuously ran this micro-optimization loop throughout the shift. As the afternoon heated up, the agent incrementally adjusted the parameters, keeping the process perfectly centered.

The quantitative results were substantial. The afternoon scrap rate spike was entirely eliminated, stabilizing at the baseline 2%. The need for manual engineering intervention was reduced by 90%, freeing the process engineers to focus on continuous improvement projects rather than firefighting daily variances. Furthermore, because the Bayesian Optimizer was constantly seeking the most efficient operating point, the system inadvertently discovered a parameter combination that reduced the overall cycle time by 1.2 seconds, resulting in a 4% increase in total daily throughput.

This case study illustrates the profound economic value of moving from static to agentic recipes. By combining the semantic reasoning of LLMs to bound the problem with the mathematical precision of ML algorithms to solve it, manufacturers can achieve a level of dynamic process control that was previously impossible, transforming their physical assets into highly adaptive, self-optimizing systems.

Key Takeaways

- **Dynamic Adaptation:** Static recipes fail under environmental and material drift. Agentic Optimizers continuously adjust parameters to maintain optimal yield despite stochastic physical conditions.

- **Separation of Concerns:** LLMs should not perform complex optimization math. They should act as orchestrators, using their semantic understanding to define the problem space and invoking specialized tools (like Bayesian Optimizers) to find the exact numerical solutions.
- **Policy-as-Code Guardrails:** Autonomous control in physical environments requires strict, deterministic safety funnels. All agent proposals must pass hard boundary checks (e.g., Pydantic validation) and physics simulations before actuation.
- **Human-in-the-Loop (HITL):** For high-risk or highly regulated processes, the agent acts as a copilot, presenting its reasoning and proposed parameters to a human operator for final approval via a secure gateway.
- **Edge Deployment:** To ensure low latency and continuous operation independent of cloud connectivity, industrial agentic loops are best deployed at the edge, integrating directly with local SCADA/MES systems via OPC UA.

Learning Objectives

- Understand the architectural shift from reactive, manual Root Cause Analysis (RCA) to autonomous, multi-modal agentic investigation in manufacturing environments.
- Design a multi-agent state machine capable of querying disparate industrial data silos (SCADA, MES, QMS, CMMS) to formulate and validate defect hypotheses.
- Implement GraphRAG and multi-modal context fusion to bridge the semantic gap between time-series anomalies, visual defect data, and unstructured maintenance logs.
- Deploy an automated Corrective and Preventive Action (CAPA) drafting pipeline that reduces Mean Time to Resolve (MTTR) while maintaining human-in-the-loop governance.

In the previous chapter, we explored how Agentic Parameter Optimizers dynamically adjust machine setpoints to prevent defects from occurring in the first place. However, manufacturing is an inherently stochastic domain. Tooling wears down, raw material batches contain hidden impurities, and environmental conditions fluctuate beyond the control limits of even the most advanced Bayesian optimizers. When these unmodeled variances align, defects occur, resulting in scrap (discarded product) or rework (product requiring additional labor to meet specifications).

When a critical defect is detected, the immediate operational priority shifts from optimization to investigation. The factory must answer a critical sequence of questions: What exactly went wrong? Why did it go wrong? How do we ensure it never happens again? Historically, this process—Root Cause Analysis (RCA)—has been one of the most labor-intensive, cognitively demanding, and time-consuming activities in industrial engineering. It relies heavily on the tribal knowledge of senior engineers and their ability to manually correlate data across highly fragmented software systems.

This chapter introduces the **Scrap & Rework Root-Cause Agent**, a specialized cognitive architecture designed to autonomously execute the investigative phases of RCA. By orchestrating multi-modal Large Language Models (LLMs) with deterministic data-querying tools, this agentic pattern transforms RCA from a days-long forensic exercise into a near-instantaneous, data-driven diagnostic process.

1.13 The Anatomy of Agentic Root-Cause Analysis

To understand the necessity of an agentic approach to RCA, we must first examine the structural deficiencies of traditional manufacturing diagnostics. The industry standard for problem-solving is the "8D" (Eight Disciplines) methodology or the simpler "5 Whys" framework. Both require engineers to iteratively ask questions to drill down from a symptom to a fundamental cause.

The bottleneck in this process is not the framework itself, but the data architecture of the modern factory. Manufacturing data is notoriously siloed into distinct layers of the ISA-95 automation pyramid, each holding a fundamentally different data type with its own blind spot, as summarized in Table~1.1.

Table 1.1: Siloed data layers of the ISA-95 automation pyramid

Dimension	SCADA/Historian	MES	ERP/QMS/CMMS
ISA-95 Level	Level 1/2	Level 3	Level 4
Data type	Structured high-frequency time-series	Relational records	Unstructured/semi-structured text
Example content	Temperatures, pressures, vibrations	Batches, work orders, routing, operator	Defect reports, SOPs, technician notes
Key blind spot	Lacks semantic context (*what* was made)	Lacks micro-second physical machine state	—

When a defect occurs, a human engineer must act as the manual integration layer between these systems. They look at a picture of a scratched part, log into the MES to find the batch number, log into the Historian to pull the time-series data for that specific batch window, export it to Excel, look for anomalies, and then search the CMMS to see if maintenance was recently performed on that machine. This process can take anywhere from 4 to 72 hours, during which the production line may be halted, or worse, continuing to produce scrap.

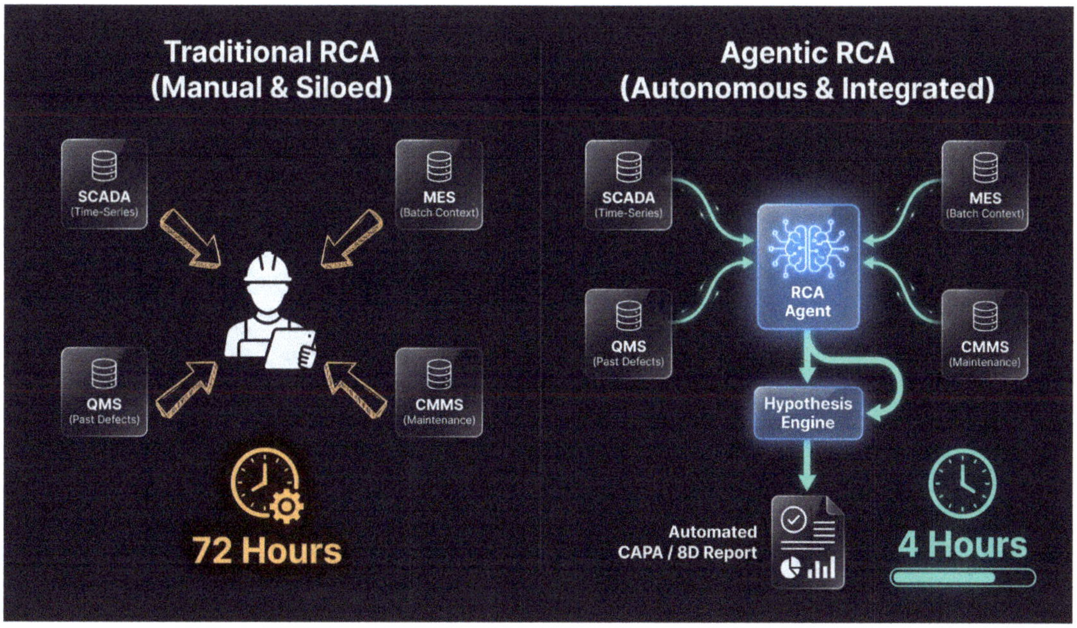

Figure 1.11: The paradigm shift from manual, siloed data gathering to autonomous, agent-driven Root Cause Analysis. The agent acts as the universal semantic bridge across the ISA-95 stack.

Traditional Machine Learning (ML) and anomaly detection algorithms fail to solve this problem comprehensively. An Isolation Forest or Autoencoder can easily flag the pressure spike in the Historian data, but it cannot read the technician's note in the CMMS, nor can it reason that the scored shaft mentioned last week is the physical cause of the pressure spike today. Traditional ML provides *detection*; it does not provide *explanation*.

The Scrap & Rework Root-Cause Agent solves this by leveraging the reasoning capabilities of frontier LLMs (like Gemini 3.5 Pro) combined with the tool-use patterns discussed in Chapter 4. The agent is not a single script, but a cognitive architecture that mimics the investigative process of a senior engineer. It can execute a "1000 Whys" analysis in seconds.

When triggered by a defect, the agent dynamically generates SQL queries to the MES to establish context, generates Flux/PromQL queries to the Historian to extract the relevant time-series window, and uses Retrieval-Augmented Generation (RAG) to search the CMMS and QMS for semantic clues. Most importantly, it uses its internal world model of physics and manufacturing principles to correlate these disparate data points. It understands the semantic bridge: it knows that a "pressure spike" (time-series anomaly) combined with a "scored shaft" (textual maintenance log) is a highly probable cause for a "dimensional variance" (quality defect). This ability to fuse multi-modal data and reason across domains is the defining characteristic of agentic RCA.

1.14 Architectural Blueprint for the RCA Agent

Building a robust RCA agent requires moving beyond simple prompt chaining into a stateful, multi-agent orchestration framework. Because the investigation process is inherently non-linear—a clue found in the maintenance log might require the agent to go back and query a different sensor in the historian—we must use a graph-based architecture. In this section, we will design the system using LangGraph, though the principles apply equally to CrewAI or the Google Agent Development Kit (ADK).

The architecture is built around a central `InvestigationState` that acts as the shared memory for the investigation. The workflow is divided into specialized nodes, often implemented as sub-agents with specific toolsets.

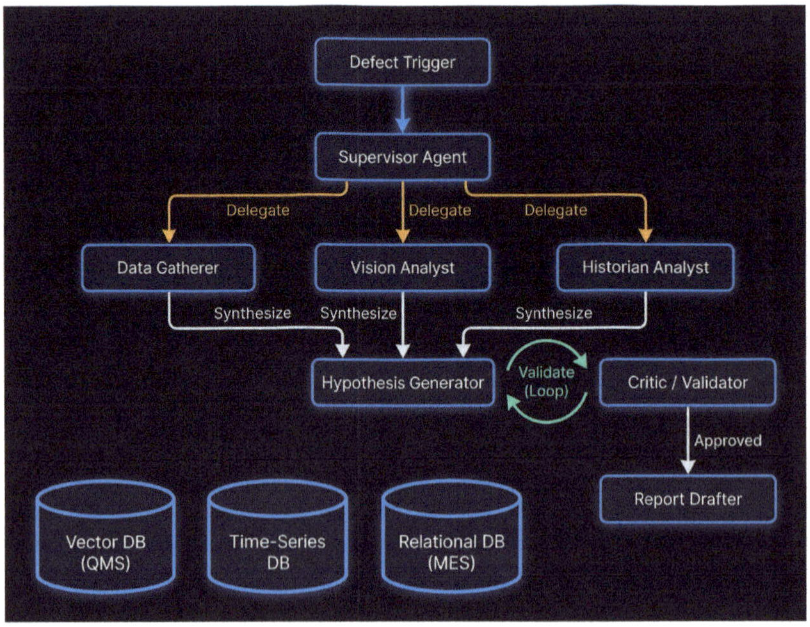

Figure 1.12: Multi-Agent RCA Architecture. The system utilizes a Supervisor-Worker pattern where specialized agents gather multi-modal data, generate hypotheses, and rigorously validate them against historical facts before drafting a final report.

The core components of the architecture include:

1. **The Trigger Node:** The entry point. It receives the initial defect payload, which might include an image from a Vision AI system, a batch ID, and a description of the failure mode (e.g., "Out of tolerance on diameter").
2. **The Context Gatherer (MES/ERP):** An agent equipped with Text-to-SQL tools. It queries the relational databases to find out exactly which machine produced the part, what time it was produced, what raw material lot was used, and who the operator was.
3. **The Historian Analyst:** An agent equipped with tools to query time-series databases (like InfluxDB or OSIsoft PI). It uses the time window provided by the Context Gatherer to pull sensor data. Crucially, it uses statistical tools (like Matrix Profile or simple Z-score thresholding) to identify anomalies in that window, rather than feeding raw time-series data directly into the LLM context window (which is inefficient and prone to hallucination).
4. **The Semantic Searcher (QMS/CMMS):** An agent equipped with vector search tools. It queries the maintenance logs and past defect reports using RAG to find historical precedents or recent maintenance activities on the specific machine.

5. **The Hypothesis Generator & Critic Loop:** The cognitive core of the system. The Generator synthesizes all gathered data to propose potential root causes. The Critic evaluates these hypotheses against the laws of physics and the gathered evidence. If a hypothesis is weak, the Critic forces the system to loop back and gather more specific data.

Below is a foundational implementation of this architecture using LangGraph and Pydantic to enforce strict state management.

Listing 1.9: Defining the LangGraph State and Hypothesis Generation Node for the RCA Agent

```
from typing import TypedDict, List, Dict, Any, Optional
from pydantic import BaseModel, Field
from langgraph.graph import StateGraph, END
from langchain_core.messages import HumanMessage, SystemMessage
from langchain_google_genai import ChatGoogleGenerativeAI

# 1. Define the strict state schema for the investigation
class InvestigationState(TypedDict):
    defect_id: str
    initial_symptom: str
    defect_image_url: Optional[str]
    mes_context: Dict[str, Any]          # Batch, Machine, Time, Material Lot
    historian_anomalies: List[Dict]      # Identified sensor spikes/drops
    maintenance_logs: List[str]          # Relevant CMMS entries
    generated_hypotheses: List[str]
    validated_root_cause: Optional[str]
    capa_report: Optional[str]
    iteration_count: int

# 2. Define the Hypothesis Generator Node
def generate_hypotheses(state: InvestigationState) -> InvestigationState:
    """Synthesizes multi-modal data to propose root causes."""

    llm = ChatGoogleGenerativeAI(model="gemini-3.5-pro", temperature=0.2)

    # Construct a highly detailed prompt injecting all gathered context
    prompt = f"""
    You are a Senior Manufacturing Reliability Engineer conducting a Root Cause Analysis.

    DEFECT SYMPTOM: {state['initial_symptom']}
    MACHINE CONTEXT: {state['mes_context']}

    TIME-SERIES ANOMALIES DETECTED DURING BATCH:
    {state['historian_anomalies']}

    RECENT MAINTENANCE LOGS FOR THIS MACHINE:
    {state['maintenance_logs']}

    Based on the principles of mechanical engineering and manufacturing physics,
    synthesize this data and generate the top 3 most probable root causes for this defect.
    Explain the causal chain for each hypothesis.
```

```
    """

    messages = [
        SystemMessage(content="You are an expert industrial diagnostic AI."),
        HumanMessage(content=prompt)
    ]

    # If a defect image exists, we would append it to the HumanMessage here
    # leveraging Gemini's native multimodal capabilities.

    response = llm.invoke(messages)

    # In a production system, we would use structured output (e.g., .with_structured_output)
    # to parse the response into a strict list of hypothesis strings.
    # For simplicity, we append the raw text here.
    state['generated_hypotheses'] = [response.content]
    state['iteration_count'] += 1

    return state

# 3. Define the Critic Node (Validation)
def validate_hypotheses(state: InvestigationState) -> InvestigationState:
    """Evaluates hypotheses and decides if more data is needed."""
    # Logic to evaluate the hypothesis against known constraints.
    # If valid, sets state['validated_root_cause'].
    # If invalid and iteration_count < MAX, requests more data.
    pass

# 4. Routing Logic
def route_after_validation(state: InvestigationState) -> str:
    if state.get('validated_root_cause'):
        return "draft_report"
    elif state['iteration_count'] >= 3:
        return "draft_report" # Force exit if stuck in a loop
    else:
        return "gather_more_data" # Loop back to specific data gathering nodes

# 5. Build the Graph
workflow = StateGraph(InvestigationState)
# ... (add nodes for data gathering, generation, validation, reporting) ...
workflow.add_node("generate_hypotheses", generate_hypotheses)
workflow.add_node("validate_hypotheses", validate_hypotheses)

workflow.add_edge("generate_hypotheses", "validate_hypotheses")
workflow.add_conditional_edges("validate_hypotheses", route_after_validation)
# ... (compile and run) ...
```

This architecture ensures that the LLM is not just guessing based on a text description of the problem. It is forced to ground its reasoning in the deterministic data pulled from the factory's systems of record. The `iteration_count` acts as a crucial safety mechanism, preventing infinite loops if the agent encounters a truly novel failure mode that cannot be explained by the available data. In such cases, the agent gracefully degrades, drafting a report that summarizes the anomalies it *did* find and handing the investigation over to a human engineer, having still saved hours of data-gathering time.

1.15 Multi-Modal Defect Analysis and Knowledge Grounding

The true power of the RCA Agent is unlocked when we move beyond structured text and embrace multi-modal analysis and advanced knowledge grounding. In manufacturing, a picture is literally worth a thousand data points. A scratch on a machined surface, the discoloration of a molded plastic part, or the specific fracture pattern of a broken tool contains immense diagnostic information that is incredibly difficult to describe in text but instantly recognizable to an expert eye—or a frontier vision model.

By utilizing models like Gemini 3.5 Pro, the RCA Agent can ingest the high-resolution image from the Automated Optical Inspection (AOI) camera directly into its context window alongside the time-series and maintenance data.

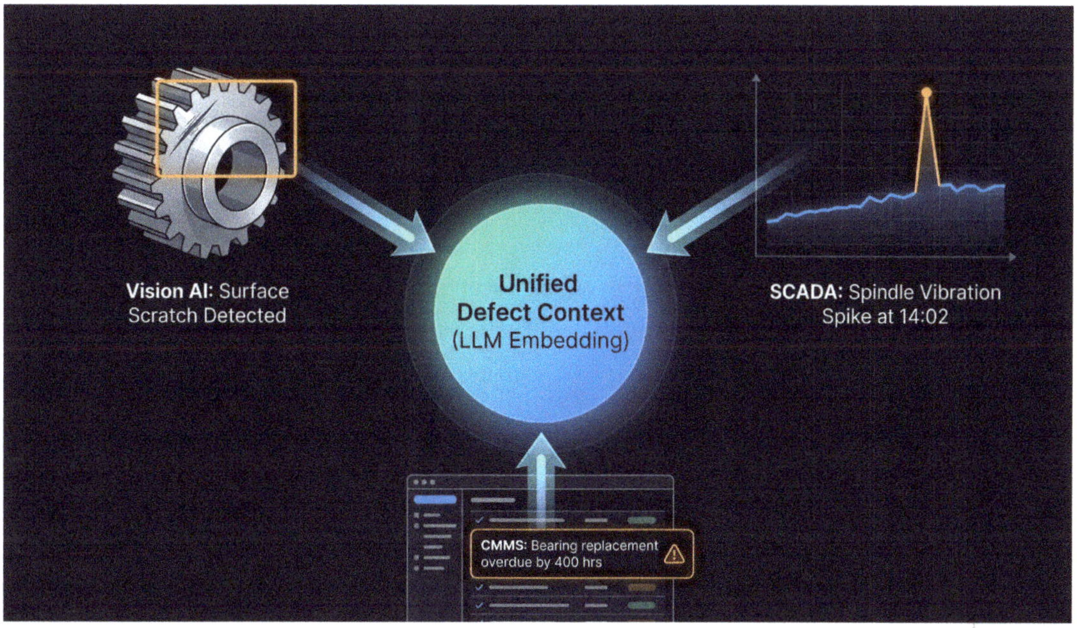

Figure 1.13: Multi-Modal Context Fusion. The agent correlates spatial anomalies (vision), temporal anomalies (time-series), and semantic context (text) to form a holistic understanding of the defect.

Consider a scenario where a CNC machine produces a part with a poor surface finish (chatter marks).

1. The Vision AI detects the chatter marks and triggers the RCA Agent, passing the image.
2. The agent analyzes the image and identifies the specific frequency and pattern of the chatter.
3. The agent queries the Historian for spindle vibration data during that specific cut.
4. The agent queries the CMMS for the tool life of the endmill used.

The LLM can now reason: "The image shows high-frequency chatter marks characteristic of tool deflection. The Historian data shows a corresponding 15% spike in spindle vibration at the exact moment of the finishing pass. The CMMS indicates this endmill was at 95% of its expected life. Conclusion: Tool wear led to deflection and vibration, causing the surface defect."

However, to make this reasoning robust across a massive enterprise, standard RAG (Retrieval-Augmented Generation) is often insufficient. Standard vector databases retrieve text based on semantic similarity. If the agent searches for "chatter marks," it might retrieve a document about chatter on Machine A, even though the current defect is on Machine B, which has a completely different kinematic structure.

To solve this, advanced RCA Agents utilize **GraphRAG** (Graph Retrieval-Augmented Generation) built on a Manufacturing Ontology. A knowledge graph maps the physical and logical relationships of the factory: `[Machine_B] -HAS_COMPONENT-> [Spindle_4] -MAINTAINED_BY-> [Tech_Smith]`.

When the agent investigates a defect on Machine B, it doesn't just do a blind vector search. It traverses the knowledge graph. It pulls the maintenance history specifically linked to `[Spindle_4]` and the SOPs linked to `[Machine_B]`. This topological grounding prevents the LLM from hallucinating connections between unrelated pieces of equipment. It provides the agent with "Episodic Memory"—the ability to recall that "the last time Spindle 4 exhibited this specific vibration profile, it was due to a loose collet nut, as documented in CAPA-2024-081." By combining multi-modal perception with graph-grounded memory, the agent achieves a level of diagnostic accuracy that rivals human domain experts.

1.16 Production Case Study: Semiconductor Wafer Fabrication

To illustrate the profound economic and operational impact of the Scrap & Rework Root-Cause Agent, we examine its deployment in one of the most complex manufacturing environments on earth: a Semiconductor Wafer Fabrication facility (Fab).

In semiconductor manufacturing, a single silicon wafer undergoes hundreds of sequential process steps over several months. If a defect is introduced early in the process but not detected until metrology checks weeks later, the financial loss is catastrophic—often tens of thousands of dollars per scrapped wafer, plus the lost opportunity cost of the fab's capacity. Rapid, accurate RCA is not just an optimization; it is critical to the fab's survival.

This specific case study focuses on the Chemical Mechanical Planarization (CMP) process. CMP uses chemical slurries and mechanical polishing pads to flatten the surface of the wafer between photolithography steps. The fab was experiencing intermittent "micro-scratching" on the oxide layer, leading to yield drops. The traditional RCA process was agonizingly slow. When metrology detected a scratch, a Yield Engineer had to manually pull the tool logs (thousands of sensor readings per second), check the slurry batch records, review the pad conditioning history, and interview the operators. The Mean Time to Resolve (MTTR) for these complex CMP defects averaged 48 hours.

The fab engineering team deployed an RCA Agent built on the Google Agent Development Kit (ADK), utilizing Gemini 3.5 Pro for its massive context window and multi-modal capabilities. The agent was integrated directly into the fab's Yield Management System (YMS).

The operational flow was as follows:

1. **Detection & Trigger:** The automated defect review scanning electron microscope (ADR-SEM) detected a cluster of micro-scratches on Wafer #742. The YMS automatically triggered the RCA Agent via API, passing the SEM image, the wafer ID, and the timestamp of the CMP step.
2. **Multi-Modal Ingestion:** The agent analyzed the SEM image. The LLM identified the morphology of the scratches as "comet tails," a specific pattern typically associated with slurry agglomeration (clumping) or pad debris.
3. **Data Orchestration:** The Supervisor Agent dispatched its workers.
   - The *MES Agent* confirmed the wafer was processed on CMP Tool 4, Platen 2, using Slurry Batch 88A.
   - The *Historian Agent* queried the time-series data for Platen 2 during the exact 90-second polishing window. It ran a statistical anomaly detection script and found a transient 5% drop in slurry flow rate, followed by a micro-spike in platen motor torque.
   - The *QMS Agent* used GraphRAG to search for historical occurrences of "comet tail scratches" linked to "flow rate anomalies" on CMP tools.
4. **Synthesis & Hypothesis:** The agent synthesized the findings. The historical QMS data revealed that a transient drop in slurry flow can cause localized drying on the pad, leading to slurry agglomeration. The agglomerated particles then cause the comet-tail scratches, which increases friction, explaining the torque spike.
5. **Root Cause Identification:** But *why* did the flow drop? The agent queried the CMMS for the slurry delivery system. It found a maintenance log from two days prior: "Replaced filter housing on Slurry Line B. Noticed slight wear on the pneumatic valve diaphragm, but did not have spare in stock. Reassembled."
6. **CAPA Generation:** The agent concluded the root cause: The worn pneumatic valve on Slurry Line B was sticking intermittently, causing transient flow drops, leading to slurry agglomeration and subsequent wafer scratching. It automatically drafted an 8D report detailing this causal chain, attached the relevant graphs and logs, and proposed a Corrective Action: "Immediately replace pneumatic valve on Slurry Line B. Update inventory min-max levels for valve diaphragms to prevent future stockouts."

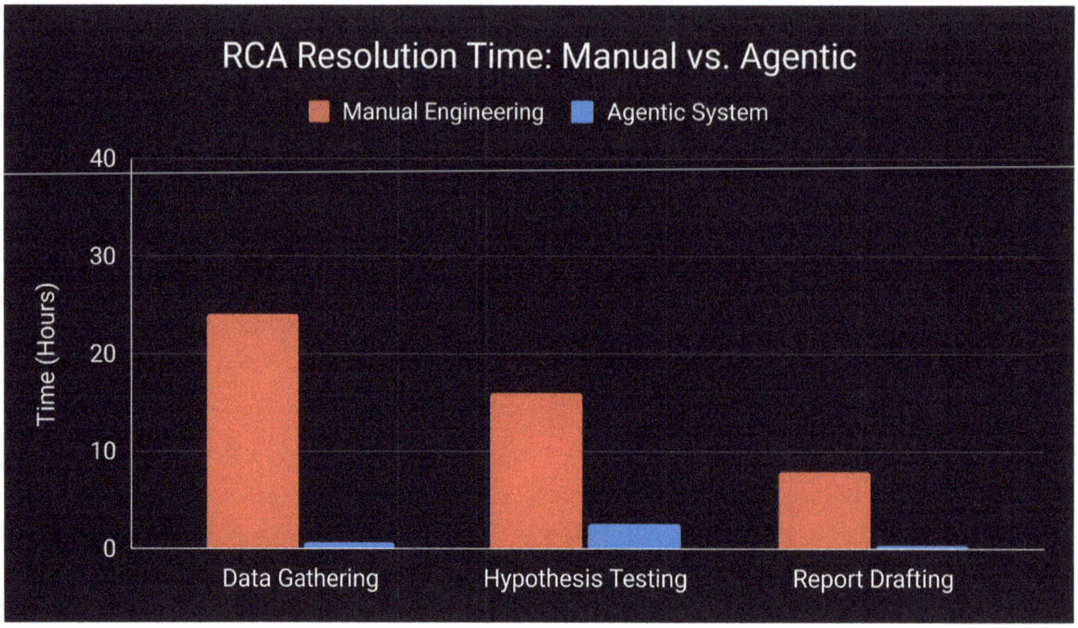

Figure 1.14: Impact of the RCA Agent on Resolution Time. The most significant time savings occur in the 'Data Gathering' phase, where the agent autonomously queries disparate silos in minutes rather than days.

The entire investigation, from the SEM image trigger to the drafted 8D report sitting in the Yield Manager's inbox, took 14 minutes. The manager reviewed the agent's logic, verified the maintenance log, and approved the CAPA. The tool was taken down, the valve was replaced, and the scratching issue was resolved.

The quantitative results of deploying this agent across the fab were staggering. The overall MTTR for complex defects was reduced from 48 hours to under 4 hours (including human review time). This rapid resolution prevented hundreds of subsequent wafers from being processed on compromised equipment, saving an estimated $2.4 million in scrap costs in the first quarter alone. Furthermore, it freed up thousands of hours of highly specialized engineering time, allowing the yield team to focus on next-generation process development rather than forensic data gathering.

This case study demonstrates that the true value of Agentic AI in manufacturing is not just in automating physical tasks, but in automating cognitive workflows. By acting as a tireless, multi-modal investigator that can instantly traverse the complex data topology of a modern factory, the RCA Agent fundamentally changes the economics of quality control.

Key Takeaways

- **Cognitive Automation over Physical Automation:** The RCA Agent automates the most time-consuming part of problem-solving: data gathering and correlation across siloed IT and OT systems (SCADA, MES, QMS, CMMS).
- **Multi-Modal Synthesis is Mandatory:** Text alone is insufficient for manufacturing

diagnostics. The ability of frontier models to ingest defect images (vision) alongside time-series anomalies and maintenance logs is critical for accurate root cause identification.

- **GraphRAG for Topological Context:** Standard vector search fails in complex factories. Using GraphRAG ensures the agent understands the physical and logical relationships between machines, parts, and historical defects, preventing hallucinated correlations.
- **Human-in-the-Loop Governance:** The agent does not execute physical repairs or unilaterally change standard operating procedures. It acts as an expert investigator, drafting comprehensive 8D/CAPA reports for human engineering approval, ensuring compliance and safety.
- **Massive MTTR Reduction:** By reducing the time required to find a root cause from days to minutes, the agent prevents subsequent products from being manufactured on compromised equipment, directly reducing scrap costs and increasing overall yield.

In the previous chapter, we explored how Agentic AI transforms quality control by autonomously investigating the root causes of scrap and rework. We established that stabilizing product quality is the foundational prerequisite for advanced manufacturing operations. However, once a process is stable and quality is assured, the operational imperative immediately shifts to efficiency. In the context of Industry 4.0 and the global push toward decarbonization, the most critical metric of efficiency is Specific Energy Consumption (SEC), commonly referred to as Energy-per-Unit (EPU).

Manufacturing accounts for roughly one-third of global energy consumption. Historically, energy was treated as a fixed overhead cost—a necessary utility to keep the machines running. Today, volatile energy markets, dynamic grid pricing, and stringent ESG (Environmental, Social, and Governance) mandates have transformed energy into a highly variable, actively managed raw material. The challenge is that traditional control systems are structurally incapable of optimizing energy consumption dynamically without risking product quality or throughput.

This chapter introduces the **Energy-per-Unit (EPU) Minimizer Agent**, a specialized architectural pattern that continuously orchestrates machine setpoints, production schedules, and grid interactions to minimize the energy required to produce a single unit of output. By bridging the semantic gap between Operational Technology (OT) and Information Technology (IT), this agentic pattern moves factories from reactive energy monitoring to proactive, autonomous energy optimization.

Learning Objectives

- Understand the thermodynamic and control-theory limitations of static setpoints and traditional Advanced Process Control (APC) in energy optimization.
- Architect an EPU Minimizer Agent that safely integrates real-time OT data (SCADA/Historian), IT context (MES/ERP), and external signals (Grid APIs).
- Implement a "Surrogate Model" pattern to allow the LLM agent to simulate the

quality and throughput impacts of proposed energy-saving actions before execution.
- Design multi-agent negotiation protocols where the EPU Agent collaborates with Production Scheduling Agents to leverage the factory as a flexible grid asset.
- Deploy safety guardrails and action-space clamping to ensure autonomous setpoint adjustments never violate mechanical limits or safety standards.

1.17 The Physics and Economics of Energy-per-Unit (EPU) Optimization

To understand why an Agentic AI is required for EPU optimization, we must first examine the physics of manufacturing energy consumption and the limitations of legacy control systems.

In any industrial process—whether it is plastic injection molding, metal machining, chemical distillation, or food extrusion—energy is consumed in two primary categories: **Baseline Energy** and **Variable Energy**. Baseline energy is the power required simply to keep the machine on, maintain operating temperatures, and run auxiliary systems (like coolant pumps or compressed air). Variable energy is the power directly applied to transform the material (e.g., the torque applied by a spindle to cut metal, or the heat applied to melt resin).

The fundamental equation for Energy-per-Unit is:

$$EPU = \frac{E_{baseline} + E_{variable}}{Throughput}$$

Because $E_{baseline}$ is often a massive fixed cost, the traditional approach to minimizing EPU has been to maximize throughput. "Run the machine as fast as possible" dilutes the baseline energy over a larger number of units. However, this brute-force approach ignores the non-linear realities of thermodynamics and tool wear. Pushing a machine faster often causes $E_{variable}$ to spike exponentially due to increased friction, heat generation, and mechanical resistance. Furthermore, running at maximum speed frequently pushes the process to the edge of its quality control limits, leading to scrap. As we established in the previous chapter, the energy embedded in a scrapped part is entirely wasted, catastrophically destroying the true EPU metric.

1.17.1 The Failure of Static Setpoints and Traditional APC

For decades, factories have relied on Proportional-Integral-Derivative (PID) controllers to maintain static setpoints (e.g., "Keep the chiller water at exactly 45°F" or "Maintain the oven at 350°C"). These setpoints are typically established during the initial commissioning of the machine and are rarely changed. They are designed for the "worst-case scenario"—ensuring the machine can handle the heaviest load on the hottest day of the year.

Consequently, for 95% of the production cycle, the machine is over-consuming energy. It is cooling water more than necessary, heating air more than required, and running pumps at 100% speed when 60% would suffice.

Advanced Process Control (APC) and Model Predictive Control (MPC) attempted to solve this by using mathematical models to adjust setpoints dynamically. While MPC is highly effective for continuous, steady-state chemical processes, it struggles profoundly in discrete manufacturing or highly variable batch processes. Why? Because MPC lacks *semantic context*.

An MPC controller only understands numerical arrays. It does not know that "Today we are running Product SKU-88, which has a higher tolerance for temperature variance than SKU-99." It does not know that "The production schedule shows a 2-hour planned maintenance window at 2:00 PM, so we can let the thermal mass of the oven drift downward starting at 1:30 PM." It does not know that "The local energy grid has issued a Demand Response event, and electricity prices will spike by 400% in 15 minutes."

1.17.2 The Agentic Paradigm Shift

The EPU Minimizer Agent solves the semantic gap. By utilizing Large Language Models (LLMs) equipped with tool-calling capabilities, the agent can ingest, parse, and reason over unstructured and structured data from across the enterprise. It reads the Manufacturing Execution System (MES) to understand the schedule and product requirements. It queries external APIs to monitor grid pricing and weather forecasts (which impact HVAC and cooling tower efficiency). It reads the real-time SCADA data to understand the current physical state of the machine.

Most importantly, the agent can reason about the *trade-offs* between energy, quality, and throughput. It acts as a continuous optimizer, asking itself: "Given the current product, the upcoming schedule, and the current price of electricity, what is the absolute minimum energy state this machine can safely occupy right now without violating quality constraints?"

This is not a simple rule-based system. The agent dynamically generates hypotheses, simulates the outcomes of those hypotheses using digital twins, and then executes the optimal setpoint changes via edge controllers. This continuous, context-aware optimization loop routinely reduces EPU by 10% to 30% in mature manufacturing facilities that have already exhausted traditional efficiency measures.

1.18 Architecture of the EPU Minimizer Agent

Architecting an agent that has write-access to physical industrial machinery requires an extreme focus on safety, determinism, and latency. The EPU Minimizer Agent cannot simply "guess" a setpoint and push it to a Programmable Logic Controller (PLC). The architecture must enforce strict boundaries between the non-deterministic reasoning of the LLM and the deterministic execution of the OT layer.

The architecture consists of four primary layers: Context Ingestion, Agentic Reasoning, Surrogate Simulation, and Guardrailed Execution.

1.18.1 1. Context Ingestion Layer

The agent operates on a continuous loop (e.g., every 5 minutes). At the start of each cycle, it gathers the current state of the world using specific tools:

- **OT Telemetry Tool:** Queries the Data Historian (e.g., OSIsoft PI, Google Cloud Manufacturing Data Engine) for the last 15 minutes of sensor data (temperatures, pressures, flow rates, power draw).

- **MES Context Tool:** Queries the MES to determine the current active work order, the specific SKU being produced, and the target throughput rate.
- **Grid/Environment Tool:** Queries external APIs for real-time electricity spot prices, upcoming Demand Response events, and ambient temperature/humidity (crucial for thermodynamic calculations).

1.18.2 2. The Surrogate Model Pattern

This is the most critical architectural pattern for industrial agents. An LLM, even a frontier model like Gemini 3.5 Pro, cannot natively calculate the exact thermodynamic impact of changing a chiller setpoint from 45°F to 48°F on the crystalline structure of a specific polymer. It lacks the deterministic physics engine.

Instead, we provide the agent with a **Surrogate Model Tool**. A surrogate model is a fast, lightweight machine learning model (often a Random Forest, Gradient Boosted Tree, or a small neural network) trained on historical factory data. It takes machine setpoints as inputs and predicts two things: (1) The resulting Energy Consumption, and (2) The resulting Product Quality (e.g., dimensional variance).

The LLM acts as the *search heuristic*. It uses its semantic understanding to propose intelligent setpoint changes, but it *must* pass those proposals through the Surrogate Model to get a deterministic prediction before proceeding.

1.18.3 3. Agentic Reasoning and Planning

With the context gathered and the Surrogate Model available, the agent enters its reasoning phase. We implement this using a state graph framework like LangGraph. The agent evaluates the current EPU against the theoretical minimum. If the grid price is high, it might propose relaxing a temperature constraint slightly (within the bounds allowed by the specific SKU's recipe). It queries the Surrogate Model: "If I raise the cooling water temp by 2 degrees, what is the predicted energy savings, and does the predicted quality metric stay within the Six Sigma control limits?"

If the Surrogate Model predicts a quality failure, the agent reflects, adjusts its proposal, and queries again. This is a classic Critic-Refiner loop, grounded in deterministic physics.

1.18.4 4. Guardrailed Execution Layer

Once the agent finds an optimal set of parameters that reduce energy without violating quality, it attempts to execute the change. However, the agent does not write directly to the PLC. It writes to a **Guardrail Gateway**. This is a hard-coded, deterministic software layer that enforces absolute physical limits. For example, a rule might state: "Regardless of what the AI requests, the chiller setpoint can never exceed 55°F, and it can never change by more than 1°F per minute." If the agent's request violates the guardrail, the request is clamped or rejected, and an error is returned to the agent for self-correction.

1.18.5 Implementation: LangGraph EPU Optimization Loop

The following code demonstrates the core logic of the EPU Minimizer Agent using LangGraph and the Google Agent Development Kit (ADK) paradigms. It illustrates the interaction between the LLM planner, the surrogate model, and the safety guardrails.

Listing 1.10: LangGraph Implementation of the EPU Minimizer Loop

```
import os
from typing import TypedDict, List, Dict, Any
from langgraph.graph import StateGraph, END
from langchain_google_genai import ChatGoogleGenerativeAI
from langchain_core.messages import SystemMessage, HumanMessage

# 1. Define the State Schema
class EPUState(TypedDict):
    current_telemetry: Dict[str, float]
    mes_context: Dict[str, Any]
    grid_price: float
    proposed_setpoints: Dict[str, float]
    surrogate_prediction: Dict[str, float]
    execution_status: str
    iteration_count: int

# 2. Define the Tools (Mocked for illustration)
def query_surrogate_model(setpoints: Dict[str, float], context: Dict[str, Any]) -> Dict[str, float]:
    """Simulates the physical process. Returns predicted energy and quality."""
    # In production, this calls a deployed ML model (e.g., Vertex AI Endpoint)
    # Mock logic: higher temp = lower energy, but higher quality variance
    energy_draw = 100 - (setpoints.get('chiller_temp', 45) - 45) * 2
    quality_variance = 0.01 + (setpoints.get('chiller_temp', 45) - 45) * 0.005
    return {"predicted_kw": energy_draw, "predicted_variance": quality_variance}

def enforce_guardrails(setpoints: Dict[str, float]) -> bool:
    """Hard-coded safety limits."""
    if setpoints.get('chiller_temp', 45) > 52.0:
        return False
    return True

# 3. Define the Nodes
def ingest_context(state: EPUState) -> EPUState:
    """Gathers data from OT, IT, and Grid."""
    # Mocking API calls to Historian, MES, and Grid
    state['current_telemetry'] = {'chiller_temp': 45.0, 'pump_speed': 1800}
    state['mes_context'] = {'sku': 'Polymer-A', 'max_variance': 0.04}
    state['grid_price'] = 0.15 # $ per kWh
    state['iteration_count'] = 0
    return state

def plan_optimization(state: EPUState) -> EPUState:
    """LLM proposes new setpoints based on context."""
    llm = ChatGoogleGenerativeAI(model="gemini-3.5-pro", temperature=0.1)

    prompt = f"""
    You are an Energy Optimization Agent for an injection molding line.
```

```
    Current State: {state['current_telemetry']}
    Product Context: {state['mes_context']}
    Current Grid Price: ${state['grid_price']}/kWh

    Your goal is to minimize energy consumption. Propose a new 'chiller_temp'.
    The current max allowed variance for this SKU is {state['mes_context']['max_variance']}.
    Respond ONLY with a JSON object containing the proposed setpoints.
    """

    # In production, use structured output. Mocking the LLM response parsing here.
    # Assume the LLM reasons that since grid price is high, it should raise temp.
    proposed_temp = state['current_telemetry']['chiller_temp'] + 2.0
    state['proposed_setpoints'] = {'chiller_temp': proposed_temp}
    state['iteration_count'] += 1
    return state

def simulate_and_evaluate(state: EPUState) -> EPUState:
    """Queries the surrogate model and evaluates the result."""
    prediction = query_surrogate_model(state['proposed_setpoints'], state['mes_context'])
    state['surrogate_prediction'] = prediction
    return state

def route_evaluation(state: EPUState) -> str:
    """Decides whether to execute, replan, or abort."""
    pred_variance = state['surrogate_prediction']['predicted_variance']
    max_variance = state['mes_context']['max_variance']

    if pred_variance <= max_variance:
        return "execute_change"
    elif state['iteration_count'] < 3:
        return "plan_optimization" # Try again with more conservative setpoints
    else:
        return "abort" # Cannot find a safe optimization

def execute_change(state: EPUState) -> EPUState:
    """Passes setpoints through guardrails and writes to PLC."""
    if enforce_guardrails(state['proposed_setpoints']):
        # Write to PLC via OPC-UA or MQTT
        state['execution_status'] = "Success: Setpoints updated."
    else:
        state['execution_status'] = "Failed: Guardrail violation."
    return state

# 4. Build the Graph
workflow = StateGraph(EPUState)
workflow.add_node("ingest_context", ingest_context)
workflow.add_node("plan_optimization", plan_optimization)
workflow.add_node("simulate_and_evaluate", simulate_and_evaluate)
workflow.add_node("execute_change", execute_change)
```

```
workflow.add_edge("ingest_context", "plan_optimization")
workflow.add_edge("plan_optimization", "simulate_and_evaluate")
workflow.add_conditional_edges("simulate_and_evaluate", route_evaluation, {
    "execute_change": "execute_change",
    "plan_optimization": "plan_optimization",
    "abort": END
})
workflow.add_edge("execute_change", END)

# Compile the agent
epu_agent = workflow.compile()
```

This code illustrates the fundamental safety mechanism: the LLM never touches the PLC directly. It proposes, the Surrogate Model predicts, the routing logic verifies against MES constraints, and the Guardrail function verifies against physical limits. This multi-layered defense-in-depth is mandatory for any agentic system operating in a physical industrial environment.

1.19 Multi-Agent Coordination for Grid-Aware Manufacturing

While a single EPU Minimizer Agent optimizing a specific machine is powerful, the true transformative potential of Agentic AI emerges when we scale this to a multi-agent system. A modern factory is not a collection of isolated machines; it is a complex, interconnected ecosystem. Optimizing one machine in isolation can inadvertently cause bottlenecks or energy spikes elsewhere.

To achieve facility-wide optimization, we deploy a **Multi-Agent Mesh Architecture**. In this setup, individual Machine Agents (like the EPU Minimizer we just built) collaborate with higher-level orchestrators. The most critical collaboration occurs between the EPU Agents, the **Production Scheduling Agent (S&OP)**, and the **Grid Interaction Agent**.

1.19.1 The Factory as a Flexible Grid Asset

Historically, factories were inflexible consumers of electricity. They drew power when they needed it, regardless of the strain on the local utility grid. Today, with the rise of renewable energy (solar and wind), grid supply is highly variable. Utilities heavily incentivize industrial consumers to become "flexible assets"—reducing consumption during peak demand (Demand Response) or absorbing excess power when renewables overproduce.

The Grid Interaction Agent continuously monitors wholesale electricity markets, weather forecasts, and utility API signals. When it detects an impending price spike or a Demand Response event, it broadcasts a signal to the factory's agentic mesh: "Energy costs will increase by 300% between 2:00 PM and 4:00 PM. Requesting facility-wide load shedding."

1.19.2 Agentic Negotiation and Load Shifting

Upon receiving the grid signal, the agents engage in a rapid negotiation protocol (often utilizing the Agent-to-Agent or A2A protocol discussed in Chapter 65).

1. **The S&OP Agent** evaluates the production schedule. It identifies that Line 1 is producing a critical, high-margin order that cannot be delayed. However, Line 2 is producing inventory for safety stock, and there is a 4-hour buffer in the warehouse. 2. **The S&OP Agent** instructs the EPU Agent on Line 2 to minimize power consumption aggressively, even if it means reducing throughput by 50% during the 2-hour window. 3. **The EPU Agent** on the facility's central chiller plant receives the signal. It reasons: "If I drop the chilled water setpoint from 45°F to 38°F right now (at 1:00 PM when power is cheap), I can 'super-cool' the 10,000-gallon buffer tank. This tank will act as a thermal battery. At 2:00 PM, when the price spikes, I can turn off the main compressors entirely and let the factory run off the thermal inertia of the super-cooled water for two hours."

This level of dynamic, cross-domain reasoning is impossible with traditional automation. It requires the semantic understanding of inventory buffers, thermal physics, and economic trade-offs—exactly the capabilities that frontier LLMs provide when properly grounded in factory data. By orchestrating these micro-adjustments across hundreds of assets, the multi-agent system can shed megawatts of load seamlessly, turning energy flexibility into a massive revenue stream for the manufacturer.

1.20 Production Case Study: Continuous Process Optimization in a Cement Plant

To illustrate the profound impact of the EPU Minimizer Agent in a heavy industrial setting, we examine its deployment in a cement manufacturing facility. Cement production is one of the most energy-intensive processes on the planet, accounting for roughly 7% of global CO2 emissions. The core of the process is the rotary kiln—a massive steel tube, often 15 feet in diameter and 300 feet long, rotating slowly while a massive burner heats the internal temperature to over 1450°C (2640°F) to calcine limestone into clinker.

The energy dynamics of a cement kiln are notoriously difficult to optimize. The process is highly non-linear, with massive thermal inertia (changes made at the burner take hours to fully manifest at the other end of the kiln). The fuel mix is often highly variable, combining pulverized coal, natural gas, and alternative fuels like biomass or shredded tires, each with different calorific values and combustion profiles.

Historically, highly experienced human operators controlled the kiln. They relied on intuition, visual inspection of the flame via cameras, and a few key sensor readings (backend temperature, NOx emissions, oxygen levels) to adjust the fuel feed rate and the Induced Draft (ID) fan speed. Because the cost of a kiln "going cold" or producing off-spec clinker (high free lime content) is catastrophic, operators naturally ran the kiln "hot"—using significantly more fuel than theoretically necessary to maintain a wide safety margin.

The facility deployed an EPU Minimizer Agent built on the Google Agent Development Kit (ADK), utilizing Gemini 3.5 Pro for reasoning and Vertex AI for the Surrogate Model.

1.20.1 The Agentic Implementation

1. **The Surrogate Model:** The engineering team trained a deep neural network on three years of historical kiln data. The model took 40 inputs (feed rate, fuel mix ratios, ambient humidity, ID fan speed, rotation speed) and predicted the resulting backend temperature, NOx emissions,

and the estimated free lime content of the clinker 2 hours into the future. 2. **The Agentic Loop:** Every 15 minutes, the EPU Agent ingested the current state. Its prompt instructed it to: "Minimize total fuel cost (prioritizing cheaper alternative fuels) while ensuring predicted NOx remains below environmental limits and predicted free lime remains below 1.5%." 3. **Edge Deployment:** Because internet connectivity in heavy industrial plants can be intermittent, and latency is critical, the agent's execution layer and the Surrogate Model were deployed on an edge server directly connected to the plant's SCADA network via OPC-UA. The LLM reasoning occurred in the cloud, but the deterministic safety checks and PLC writes occurred locally. If the cloud connection dropped, the edge system automatically reverted to the last known safe setpoints or handed control back to the human operator.

1.20.2 Operational Results and ROI

The agent's ability to continuously simulate thousands of parameter combinations allowed it to find optimization pockets that human operators could not see. For example, the agent discovered that by slightly increasing the kiln rotation speed while simultaneously reducing the ID fan speed and increasing the ratio of biomass fuel, it could maintain the required calcination temperature while reducing overall thermal energy input.

The agent operated with a precision that eliminated the need for the massive "safety margins" human operators relied upon. It rode the edge of the optimal thermal profile, constantly making micro-adjustments to account for variations in the raw material feed and ambient weather conditions.

The quantitative results were transformative:

- **Thermal EPU Reduction:** The specific heat consumption (megajoules per ton of clinker) was reduced by 8.4%.
- **Alternative Fuel Utilization:** The agent safely increased the substitution rate of cheaper, lower-carbon alternative fuels by 12%, as it could dynamically adjust the draft fan to ensure complete combustion despite the variable calorific value of the waste fuels.
- **Financial Impact:** For a kiln producing 5,000 tons of clinker per day, the 8.4% reduction in thermal energy translated to an annualized saving of $1.8 million in fuel costs.
- **Emissions Reduction:** The optimization directly resulted in a reduction of 22,000 tons of CO_2 emissions annually per kiln, significantly advancing the company's ESG targets and reducing carbon tax liabilities.

Crucially, the human operators were not replaced; they were elevated. Instead of manually tweaking dials every 10 minutes, they transitioned to a supervisory role. They monitored the agent's performance dashboards, managed the physical logistics of the alternative fuel supply chain, and focused on long-term equipment maintenance. The EPU Minimizer Agent proved that by combining the semantic reasoning of LLMs with the deterministic physics of surrogate models, heavy industry can achieve levels of efficiency previously thought impossible.

Key Takeaways

- **Semantic Context is the Key to Efficiency:** Traditional control systems fail to optimize energy because they lack awareness of production schedules, grid pricing, and product-specific tolerances. Agentic AI bridges this gap, enabling dynamic, context-aware optimization.
- **The Surrogate Model Pattern is Mandatory:** LLMs cannot reliably calculate physical thermodynamics. The agent must use a deterministic machine learning model (the Surrogate) as a tool to simulate the quality and energy impacts of its proposed actions before execution.
- **Defense-in-Depth Safety Architecture:** Autonomous agents must never write directly to a PLC without passing through a hard-coded, deterministic Guardrail Gateway that enforces absolute physical and safety limits.
- **Multi-Agent Grid Integration:** The highest ROI is achieved when EPU Agents collaborate with Scheduling and Grid Agents, allowing the factory to act as a flexible thermal battery, shedding load during price spikes and capitalizing on cheap renewable energy.
- **Massive Economic and ESG Impact:** By continuously riding the edge of optimal efficiency rather than relying on worst-case safety margins, EPU agents routinely deliver 8% to 15% reductions in energy costs, directly improving the bottom line and reducing carbon footprints.

Learning Objectives

- Understand the limitations of traditional Overall Equipment Effectiveness (OEE) tracking and the semantic gap between descriptive metrics and prescriptive action.
- Architect an Agentic OEE Coach that integrates real-time telemetry, historical maintenance logs, and unstructured equipment manuals using GraphRAG and multi-agent reasoning.
- Design human-in-the-loop (HITL) conversational interfaces that guide shop-floor operators through dynamic, context-aware troubleshooting procedures.
- Implement a LangGraph-based state machine for autonomous root-cause analysis, capable of correlating micro-stops with multi-variate environmental and process data.
- Evaluate the quantitative ROI of deploying an OEE Coach in high-speed manufacturing environments, focusing on the reduction of Mean Time to Repair (MTTR) and the capture of tribal knowledge.

1.21 The Evolution of OEE: From Lagging Metric to Autonomous Coach

Overall Equipment Effectiveness (OEE) has been the gold standard metric for manufacturing productivity since the advent of Total Productive Maintenance (TPM) in the 1960s. Calculated as the product of Availability, Performance, and Quality ($OEE = A \times P \times Q$), it provides a single, unified percentage that represents how close a manufacturing asset is to perfect production (manufacturing only good parts, as fast as possible, with no stop time). For decades, the primary goal of Manufacturing Execution Systems (MES) and Supervisory Control and Data Acquisition (SCADA) platforms has been to calculate this number accurately and display it on large, overhead dashboards on the shop floor.

However, traditional OEE tracking suffers from a fatal flaw: it is a purely descriptive, lagging indicator. A dashboard that flashes a red "62% OEE" tells the plant manager that the line is failing, but it does absolutely nothing to explain *why* it is failing, nor does it tell the operator *how* to fix it. When a machine faults, traditional systems generate an error code (e.g., "ERR-409: Servo Over-torque"). The operator must then rely on their personal experience (tribal knowledge), consult a dusty, 500-page PDF maintenance manual, or call a maintenance technician who might take 30 minutes to arrive. During this entire diagnostic period, the machine is down, and the Availability metric plummets.

The fundamental issue is a semantic gap. Manufacturing data is abundant but context-poor. A programmable logic controller (PLC) knows that a sensor voltage dropped, but it does not know that the drop was caused by a specific batch of raw materials interacting poorly with the ambient humidity, a problem that operator "John" solved three months ago by slightly increasing the tension on a specific roller. Traditional analytics and even standard machine learning models struggle to bridge this gap because the necessary context is scattered across structured time-series databases, unstructured shift logs, complex PDF manuals, and the undocumented minds of veteran operators.

Enter the Agentic OEE Coach. By leveraging the advanced reasoning, multimodal processing, and long-context capabilities of frontier models like Gemini 3.5 Pro, we can transform OEE from a passive metric into an active, autonomous collaborator. The OEE Coach does not just report the score; it continuously monitors the data streams, detects anomalies, hypothesizes root causes, retrieves the exact relevant section of the maintenance manual, cross-references past work orders, and provides the operator with a dynamic, step-by-step remediation plan.

The shift to an agentic paradigm addresses the "tribal knowledge" crisis currently plaguing the manufacturing sector. As a generation of highly experienced operators and technicians retires, they take decades of undocumented troubleshooting intuition with them. The OEE Coach acts as a centralized, continuously learning repository of this knowledge. When an operator interacts with the Coach to solve a novel problem, the agent records the successful intervention, updates its internal semantic graph, and makes that solution instantly available to every other operator across the global enterprise.

Furthermore, the OEE Coach operates in a multi-agent ecosystem. As discussed in the previous chapter, it collaborates directly with the EPU (Energy Per Unit) Minimizer Agent. If the OEE Coach determines that a machine's performance degradation is due to thermal stress, it can negotiate with the EPU Agent to adjust cooling setpoints, balancing the trade-off between energy consumption and production throughput. It also interfaces with the Sales and Operations Planning (S&OP) Agent; if the Coach predicts a catastrophic failure requiring a 12-hour downtime, it immediately alerts the S&OP Agent to reroute production orders to alternative lines, minimizing supply chain disruption. This level of interconnected, semantic reasoning is the defining characteristic of Industry 4.0 realized through Agentic AI.

1.22 Architecture of the OEE Coach Agent

Building an effective OEE Coach requires an architecture that can seamlessly blend high-frequency, deterministic machine data with low-frequency, highly unstructured human and textual data. The system must be capable of real-time anomaly detection, deep semantic retrieval, and multi-step reasoning, all while maintaining strict safety guardrails and providing sub-second response times to the operator on the floor.

The core of the architecture is a stateful cognitive graph, typically implemented using LangGraph or the Google Agent Development Kit (ADK). This graph orchestrates the flow of information between the physical machine, the enterprise IT systems, and the Large Language Model. The architecture is divided into three primary layers: the Data Ingestion and Context Layer, the Agentic Reasoning Engine, and the Action and Interface Layer.

1.22.1 The Data Ingestion and Context Layer

The foundation of the OEE Coach is its ability to perceive the state of the manufacturing environment. This requires integrating multiple disparate data sources into a unified semantic space.

1. **Real-Time Telemetry (The "What"):** High-frequency data (vibration, temperature, pressure, motor torque) is ingested from PLCs via protocols like OPC-UA or MQTT. This data is typically stored in a time-series database (e.g., InfluxDB or AVEVA PI System). Because LLMs cannot process raw, high-frequency time-series data efficiently, we employ edge-based anomaly detection models (often lightweight autoencoders or statistical process control algorithms) to monitor the streams. When an anomaly is detected, it triggers an event, summarizing the telemetry window (e.g., "Motor 3 torque spiked by 45% over 3 seconds, followed by a sudden drop in line speed").

2. **Enterprise Context (The "Who and When"):** The agent queries the MES and ERP systems to understand the current production context. What SKU is currently running? Who is the operator on shift? When was the last preventive maintenance performed? What is the current target throughput? This structured data provides the necessary constraints for the agent's reasoning.

3. **Unstructured Knowledge (The "How"):** This is where the agentic approach truly shines. We utilize a sophisticated Retrieval-Augmented Generation (RAG) pipeline—specifically GraphRAG—to ingest OEM maintenance manuals, standard operating procedures (SOPs), historical work orders from the Computerized Maintenance Management System (CMMS), and digitized shift handover logs. GraphRAG is critical here because manufacturing equipment is highly relational. A manual might state that "Valve A controls pressure to Cylinder B." GraphRAG extracts these entities and relationships, allowing the agent to traverse the physical topology of the machine when reasoning about a fault.

1.22.2 The Agentic Reasoning Engine (LangGraph Implementation)

When a machine faults or performance degrades, the state machine is triggered. The reasoning loop follows a structured Sense-Plan-Act-Reflect pattern. Below is a simplified LangGraph implementation demonstrating the core diagnostic loop.

Listing 1.11: LangGraph Implementation of the OEE Diagnostic Loop

```python
import operator
from typing import TypedDict, Annotated, List, Dict, Any
from langchain_google_genai import ChatGoogleGenerativeAI
from langgraph.graph import StateGraph, END

# 1. Define the State
class OEEState(TypedDict):
    machine_id: str
    error_code: str
    telemetry_summary: Dict[str, Any]
    mes_context: Dict[str, Any]
    retrieved_manuals: List[str]
    historical_fixes: List[str]
    hypotheses: List[str]
    recommended_action: str
    operator_feedback: str
    status: str

# 2. Initialize the LLM
llm = ChatGoogleGenerativeAI(model="gemini-3.5-pro", temperature=0.1)

# 3. Define the Nodes
def retrieve_context(state: OEEState) -> OEEState:
    """Retrieves manuals and past work orders based on error code and telemetry."""
    # Mocking vector DB retrieval
    query = f"Error {state['error_code']} on {state['machine_id']} with {state['telemetry_summary']}"
    state['retrieved_manuals'] = vector_db.similarity_search(query, filter="manuals")
    state['historical_fixes'] = vector_db.similarity_search(query, filter="work_orders")
    return state

def generate_hypotheses(state: OEEState) -> OEEState:
    """LLM generates potential root causes based on retrieved context."""
    prompt = f"""
    You are an expert OEE Diagnostic Agent.
    Machine: {state['machine_id']} | Error: {state['error_code']}
    Telemetry: {state['telemetry_summary']}
    Context: {state['mes_context']}
    Manual Excerpts: {state['retrieved_manuals']}
    Past Fixes: {state['historical_fixes']}

    Generate the top 3 most likely root causes for this failure.
    Format as a list of strings.
    """
    response = llm.invoke(prompt)
    state['hypotheses'] = parse_list(response.content)
    return state

def formulate_action_plan(state: OEEState) -> OEEState:
```

```python
    """Selects the best hypothesis and creates a step-by-step guide."""
    prompt = f"""
    Based on these hypotheses: {state['hypotheses']}, formulate a safe,
    step-by-step troubleshooting guide for the operator.
    Prioritize actions that do not require specialized maintenance tools first.
    """
    response = llm.invoke(prompt)
    state['recommended_action'] = response.content
    state['status'] = "awaiting_operator"
    return state

def process_feedback(state: OEEState) -> OEEState:
    """Evaluates operator feedback to determine if the issue is resolved."""
    if "success" in state['operator_feedback'].lower() or "fixed" in state['operator_feedback'].lower():
        state['status'] = "resolved"
        # Trigger background task to update vector DB with this successful fix
    else:
        state['status'] = "escalate"
    return state

# 4. Build the Graph
workflow = StateGraph(OEEState)
workflow.add_node("retrieve_context", retrieve_context)
workflow.add_node("generate_hypotheses", generate_hypotheses)
workflow.add_node("formulate_action_plan", formulate_action_plan)
workflow.add_node("process_feedback", process_feedback)

workflow.add_edge("retrieve_context", "generate_hypotheses")
workflow.add_edge("generate_hypotheses", "formulate_action_plan")
workflow.add_edge("formulate_action_plan", "process_feedback")

# Conditional routing based on feedback
def route_post_feedback(state: OEEState) -> str:
    if state['status'] == "resolved":
        return END
    else:
        return "escalate_to_cmms" # Hypothetical node to create a work order

workflow.add_conditional_edges("process_feedback", route_post_feedback)

# Compile the agent
oee_coach_app = workflow.compile()
```

This architecture ensures that the LLM is never guessing. Every hypothesis is grounded in the physical reality of the telemetry and the documented reality of the manuals and historical work orders. The use of Gemini 3.5 Pro is particularly advantageous here due to its massive context window, allowing the agent to ingest entire chapters of maintenance manuals and months of shift logs simultaneously without losing fidelity.

1.23 Human-Agent Collaboration on the Shop Floor

The success of the OEE Coach hinges entirely on user adoption. Manufacturing operators are notoriously skeptical of new software, especially systems that claim to "know better" than they do. If the agent acts as a black-box dictator, issuing opaque commands, it will be ignored. Therefore, the interface must be designed around the principles of Human-in-the-Loop (HITL) collaboration, transparency, and multimodal interaction.

1.23.1 Dynamic Troubleshooting vs. Static SOPs

Traditionally, when a machine faults, an operator consults a Standard Operating Procedure (SOP). SOPs are static, linear decision trees (e.g., "If light is red, check Valve A. If Valve A is open, check Sensor B"). The problem with static SOPs is that they cannot account for the infinite combinatorial complexity of a modern manufacturing line. They assume a sterile environment where only one thing goes wrong at a time.

The OEE Coach replaces static SOPs with dynamic, context-aware troubleshooting trees. When the agent formulates an action plan (as seen in the code snippet above), it tailors the steps to the exact current state of the machine. If the agent knows from the MES that the current SKU uses a highly viscous fluid, and it knows from the telemetry that the ambient temperature has dropped by 10 degrees, it will bypass the standard "check the sensor" step and immediately instruct the operator to check the fluid pre-heater, explaining *why* it is making this recommendation.

This explainability is crucial for building trust. The agent must cite its sources. When it recommends adjusting a tensioner, the UI displays a small citation link pointing to the exact paragraph in the OEM manual or the specific historical work order from six months ago where a senior technician solved a similar issue.

1.23.2 Multimodal Interaction and Vision Capabilities

The shop floor is a physical environment; text is often an insufficient medium for communication. The OEE Coach leverages the multimodal capabilities of the Gemini model family to interact with operators naturally.

Operators interact with the Coach via ruggedized tablets or voice-activated headsets. If an operator encounters a jammed mechanism, typing a description is slow and inaccurate. Instead, the operator takes a photo or records a short video of the jam using the tablet. The image is passed to Gemini 3.5 Pro, along with the current telemetry and error codes.

The vision model analyzes the image, identifies the specific mechanical components, notes the orientation of the jammed material, and correlates this visual evidence with the telemetry. The agent might respond via text-to-speech: "I see the packaging film is bunched around the secondary drive roller. The telemetry shows a slight tension loss three seconds before the jam. This indicates the roller bearing may be slipping. Please inspect the bearing for wear; I have highlighted its location on your screen."

This multimodal loop—combining visual inspection, time-series data, and semantic reasoning—replicates the diagnostic process of a master technician. Furthermore, when the operator confirms the fix, the image of the jam, the telemetry signature, and the successful resolution are embedded into the vector database. The next time a similar tension loss occurs, the agent can proactively warn the operator to check the bearing *before* the jam happens, shifting the paradigm from reactive troubleshooting to predictive prevention.

1.24 Production Case Study: High-Speed Packaging Line

To understand the quantitative impact of the OEE Coach, we examine its deployment at a global consumer packaged goods (CPG) manufacturer, specifically on a high-speed bottling and packaging line producing 600 units per minute.

The primary challenge on this line was not catastrophic machine failure (which affects Availability), but rather "micro-stops" (which destroy the Performance metric). A micro-stop is a minor jam or fault that halts the line for less than two minutes. Because they are so brief, operators simply clear the jam, hit reset, and resume production without logging a formal maintenance ticket. Traditional OEE systems recorded the downtime but provided no insight into the cause. The line was experiencing an average of 45 micro-stops per shift, resulting in a 15% loss in overall Performance.

The engineering team deployed an OEE Coach Agent to address this specific issue. The agent was granted read-only access to the high-frequency PLC data (sampled at 100Hz), the MES (for SKU and batch data), and the environmental sensors (temperature and humidity).

1.24.1 The Agentic Correlation Engine

The breakthrough occurred because the agent could maintain a continuous, high-fidelity context window across multiple disparate data streams—a task impossible for a human operator to perform in real-time.

The agent's background monitoring loop was instructed to analyze the 30 seconds of telemetry preceding every micro-stop. Over the course of two weeks, the agent processed over 1,200 micro-stop events. Using its semantic reasoning capabilities, it identified a complex, multivariate correlation that had eluded the engineering team for years.

The agent discovered that 78% of the micro-stops at the labeling station occurred only when three specific conditions were met simultaneously: 1. The line was running a specific SKU (a bottle with a highly textured surface). 2. The ambient plant humidity exceeded 65%. 3. The temperature of the adhesive applicator dropped below 142°C (even though the official SOP stated the acceptable range was 135°C - 150°C).

The agent reasoned that the high humidity was slightly altering the surface tension of the textured bottle, requiring the adhesive to be applied at a higher temperature to achieve the necessary tackiness within the millisecond contact window. When the temperature dipped to 141°C, the label would slip, triggering a vision sensor fault and halting the line.

1.24.2 Intervention and ROI

The OEE Coach did not just report this finding; it acted on it. It generated a detailed report for the engineering team, citing the specific data correlations and proposing a hypothesis based on the physics of the adhesive (retrieved from the manufacturer's spec sheet via GraphRAG).

Furthermore, it updated its own dynamic troubleshooting logic. The next time the MES scheduled that specific SKU, and the weather sensors indicated high humidity, the OEE Coach proactively alerted the operator via the tablet: "Warning: High risk of labeling micro-stops due to current humidity. Recommend increasing adhesive applicator setpoint to 145°C for this batch."

The results were immediate and measurable:

- **Micro-stop Reduction:** The frequency of micro-stops on the labeling station dropped by 82%.
- **Performance Metric Increase:** The overall Performance component of the OEE calculation increased from 75% to 88%.
- **Financial Impact:** On a line producing 600 units per minute, recovering 13% of lost production time equated to an additional 1.8 million units produced annually without adding a single hour of overtime or capital expenditure. The estimated annual savings and revenue capture exceeded $2.4 million for this single line.
- **MTTR Reduction:** For faults that did occur, the Mean Time to Repair (MTTR) dropped from an average of 14 minutes to 4 minutes, as operators no longer had to hunt for manuals or wait for maintenance technicians; the agent guided them to the root cause instantly.

The OEE Coach represents a paradigm shift in manufacturing operations. By bridging the semantic gap between machine data and human knowledge, it transforms the shop floor from a reactive environment governed by lagging metrics into a proactive, continuously learning ecosystem. It empowers operators, preserves critical tribal knowledge, and unlocks levels of efficiency that are mathematically impossible to achieve with traditional, static automation.

Key Takeaways

- **Bridge the Semantic Gap:** Traditional OEE is a lagging, descriptive metric. The Agentic OEE Coach uses LLMs to bridge the gap between deterministic machine data (telemetry) and unstructured human context (manuals, logs), providing prescriptive, actionable guidance.
- **GraphRAG is Essential for Manufacturing:** Standard vector retrieval is insufficient for complex machinery. GraphRAG allows the agent to understand the physical topology and relational dependencies of equipment components, enabling accurate root-cause analysis.
- **Dynamic Troubleshooting Replaces Static SOPs:** The agent generates context-aware, step-by-step remediation plans tailored to the exact current state of the machine, drastically reducing Mean Time to Repair (MTTR) compared to static, linear SOPs.
- **Multimodal Interaction Builds Trust:** Operators must be able to interact with the agent naturally. Utilizing vision models to analyze photos of jammed parts or broken components allows the agent to "see" the problem, replicating the diagnostic process of a human expert.
- **Capture Tribal Knowledge Automatically:** Every successful interaction between the operator and the OEE Coach is recorded and embedded back into the agent's memory, ensuring that hard-won troubleshooting experience is institutionalized and instantly available globally.

Industrial Robotics & Cells

Chapter 56
Industrial Robotics & Cells

Architecting Autonomous Perception, Reasoning, and Actuation on the Manufacturing Floor

Learning Objectives

- **Understand the paradigm shift** from deterministic, hardcoded Programmable Logic Controller (PLC) state machines to probabilistic, agentic Vision-Language-Action (VLA) orchestration in industrial environments.
- **Architect hybrid control loops** that safely partition high-latency cognitive reasoning (planning, error recovery) from low-latency deterministic execution (kinematics, safety interlocks) using edge deployments.
- **Design multi-agent coordination frameworks** for complex manufacturing cells, enabling multiple robotic arms and vision systems to collaborate on unstructured tasks without rigid pre-programming.
- **Implement human-in-the-loop (HITL) guardrails** that translate agentic intent into verifiable, auditable actions, ensuring compliance with industrial safety standards (e.g., ISO 10218).

The manufacturing floor represents the ultimate frontier for artificial intelligence: the translation of digital reasoning into physical actuation. For decades, industrial robotics has been defined by rigid determinism. Traditional robot cells—enclosed environments where robotic arms perform welding, assembly, or material handling—are governed by Programmable Logic Controllers (PLCs) and proprietary languages (such as G-code or RAPID). These systems execute highly optimized, hardcoded trajectories. They are exceptionally fast, precise, and reliable, but they are fundamentally blind and brittle. If a part arrives on a conveyor belt rotated by three degrees, or if a lighting change affects a rudimentary machine vision template, the traditional cell halts, throws an error code, and waits for human intervention. In an era shifting rapidly toward high-mix, low-volume (HMLV) manufacturing, the cost of this rigidity is no longer sustainable. Reprogramming a cell for a new product variant often takes weeks of engineering time, rendering traditional automation economically unviable for dynamic production lines.

Agentic AI introduces a profound paradigm shift to industrial automation by decoupling the *intent* of a task from its *execution*. Instead of programming explicit joint angles and waypoints, engineers can now architect systems where an autonomous agent is given a high-level goal (e.g., "Assemble the gearbox, ensuring the planetary gears are lubricated, and discard any gears with visible surface defects"). The agentic system utilizes Vision-Language-Action (VLA) models—such as the Gemini 3.5 Pro family, which natively processes multimodal inputs including continuous video feeds and 3D point clouds—to perceive the unstructured environment. It then reasons about the spatial relationships, decomposes the high-level goal into a sequence of parameterized skills (such as `pick(object_id)`, `move_to(pose)`, `insert(force_threshold)`), and orchestrates the execution of these skills through the robot's middleware. This transforms the robot from a blind execution engine into a context-aware collaborator capable of dynamic replanning and autonomous error recovery.

However, deploying non-deterministic Large Language Models (LLMs) into physical environments introduces severe architectural and safety challenges. Industrial control systems operate on strict real-time constraints; a latency of 10 milliseconds can mean the difference between a successful insertion and a shattered end-effector. Cloud-based LLM inference, which typically operates in the 500ms to 2-second latency regime, cannot be placed directly in the inner control loop. Therefore, architecting agentic cells requires a rigorous hierarchical design. The agentic layer acts as the *orchestrator* and *planner*, operating asynchronously. It generates state machines, behavior trees, or parameterized API calls that are passed down to an Edge Deployment Orchestrator. This edge layer, often running on Google Distributed Cloud (GDC) or local Kubernetes clusters, translates the agent's intent into deterministic commands for the Robot Operating System (ROS2) or the PLC. The inner loop (kinematics, collision avoidance, and safety interlocks) remains strictly deterministic and mathematically verifiable, ensuring that even if the agent hallucinates a physically impossible trajectory, the lower-level safety controllers will reject the command.

To bridge the semantic gap between natural language reasoning and physical actuation, practitioners utilize the Tool Invocation design pattern extensively. The robot's capabilities are exposed to the agent as a library of well-defined, atomic tools. By providing the agent with a comprehensive schema of these tools, along with access to real-time telemetry and visual feeds, the agent can construct complex, multi-step plans.

Listing 1.12: Example Tool Schema for an Agentic Robot Cell Orchestrator using the Google Agent Development Kit (ADK).

```python
from google.adk.tools import tool
from pydantic import BaseModel, Field
from typing import Dict, List

class PickAndPlaceParams(BaseModel):
    target_object_id: str = Field( ... , description="The semantic ID of the object detected by the vision system (e.g., 'gear_housing_A').")
    destination_pose: Dict[str, float] = Field( ... , description="The target 6-DoF pose {x, y, z, roll, pitch, yaw} in the robot's base frame.")
    grip_force_newtons: float = Field(default=20.0, description="Force to apply via the parallel gripper.")

@tool(
    name="execute_pick_and_place",
    description="Commands the robotic arm to pick up an identified object and place it at a specified coordinate. Use this after verifying the object's location via the vision tool."
)
def execute_pick_and_place(params: PickAndPlaceParams) -> str:
    """
    Translates the agent's semantic request into a ROS2 Action Goal.
    In production, this function pushes the goal to the Edge Orchestrator
    and awaits a synchronous success/failure response from the PLC.
    """
    # 1. Validate pose against safety bounding box
    if not validate_safety_bounds(params.destination_pose):
        return "ERROR: Destination pose violates cell safety boundaries."

    # 2. Dispatch to ROS2 Action Server via Edge Gateway
    response = ros2_client.send_goal(
        action_name='/manipulator/pick_place',
        target=params.target_object_id,
        pose=params.destination_pose,
        force=params.grip_force_newtons
    )
```

```

    # 3. Return deterministic outcome to the Agent for reflection
    if response.status == "SUCCESS":
        return f"Successfully placed {params.target_object_id}."
    else:
        return f"FAILURE: {response.error_code}. Agent must replan."
```

The true power of this architecture emerges during failure modes. In a traditional setup, a failed grasp results in a hard stop. In an agentic cell, the failure is returned to the orchestrator as an observation. The agent, utilizing its visual context window, can "see" that the part slipped, reflect on the failure, and autonomously generate a recovery plan—perhaps commanding the arm to re-attempt the grasp from a different approach angle or requesting a human operator's assistance if the part is damaged. This self-correction loop drastically increases the Overall Equipment Effectiveness (OEE) of the cell.

Furthermore, as manufacturing scales, single-arm cells are giving way to complex, multi-agent environments where multiple robots, automated guided vehicles (AGVs), and human workers share the same workspace. Coordinating these entities using traditional state machines leads to an exponential explosion of edge cases. By employing multi-agent collaboration frameworks (such as LangGraph or CrewAI adapted for physical systems), each robot can be instantiated as an independent agent with its own spatial memory and toolset. These agents negotiate shared workspace access, hand off parts, and dynamically balance workloads using open protocols like the Agent-to-Agent (A2A) protocol, fundamentally transforming the factory floor into a distributed, cognitive mesh.

What This Chapter Covers

To successfully architect and deploy agentic intelligence into physical industrial cells, we must bridge the gap between high-level cognitive reasoning and low-level deterministic control. The remainder of this chapter deconstructs this architecture into six deployable design patterns:

- **Robot Cell Task Planner:** Designing the core cognitive loop that translates natural language production orders into executable, parameterized skill sequences for a robotic manipulator.
- **Multi-Arm Coordination Agent:** Implementing decentralized, multi-agent negotiation protocols to manage collision avoidance, workspace sharing, and task hand-offs between multiple robots in a single cell.
- **Vision-Based Pick-Place Copilot:** Utilizing multimodal VLA models to process continuous RGB-D camera feeds, enabling semantic object recognition, defect detection, and dynamic grasp pose estimation in unstructured bins.
- **Changeover & Tooling Scheduler:** Architecting agents that autonomously manage high-mix production transitions, verifying tool-changer states, and updating kinematic parameters without manual reprogramming.
- **Human-in-Loop Safety Sentinel:** Building the critical governance layer that enforces ISO safety standards, utilizing LLM-as-a-Judge patterns to validate generated trajectories before execution, and managing seamless human-robot handovers.

- **Edge Deployment Orchestrator:** Structuring the infrastructure required to run agentic runtimes on local Kubernetes clusters (e.g., Google Distributed Cloud), ensuring the strict latency and air-gapped security requirements of Operational Technology (OT) networks are met.

Learning Objectives

- **Architect the cognitive bridge** between high-level natural language production orders and low-level deterministic robotic kinematics using Large Language Models (LLMs) and Vision-Language-Action (VLA) models.
- **Design Semantic Scene Graphs** that translate raw RGB-D point clouds into structured, queryable representations of the physical workspace, enabling spatial reasoning.
- **Implement Skill Libraries** using the Tool Invocation pattern to expose deterministic Robot Operating System (ROS2) actions as parameterized functions callable by the agentic planner.
- **Engineer closed-loop error recovery** mechanisms that utilize the Sense-Plan-Act-Reflect (SPAR) lifecycle to autonomously handle physical failures (e.g., part slippage, collisions) without human intervention.

1.25 The Architecture of Agentic Task Planning

The fundamental challenge of industrial robotics in the 21st century is the tension between precision and adaptability. Traditional robotic cells, governed by Programmable Logic Controllers (PLCs) and programmed via proprietary languages like G-code or RAPID, are marvels of deterministic engineering. They can execute a predefined trajectory with sub-millimeter accuracy, thousands of times a day, without deviation. However, this determinism is also their greatest vulnerability. A traditional robot does not "know" it is assembling a gearbox; it only knows that it must move its end-effector to coordinate (x_1, y_1, z_1), close its gripper, and move to (x_2, y_2, z_2). If the gearbox housing is shifted by two centimeters due to a conveyor belt anomaly, the robot will blindly execute its trajectory, resulting in a collision, a shattered part, and a halted production line.

In the context of High-Mix, Low-Volume (HMLV) manufacturing—where production lines must frequently switch between different product variants—the cost of manually reprogramming these rigid state machines becomes prohibitive. The **Robot Cell Task Planner** is the architectural pattern that resolves this tension. It introduces a cognitive layer that decouples the *intent* of a task from its *execution*. By leveraging Agentic AI, specifically advanced Vision-Language-Action (VLA) models, the system can perceive the unstructured physical environment, reason about spatial relationships, and autonomously generate the sequence of deterministic commands required to achieve a high-level goal.

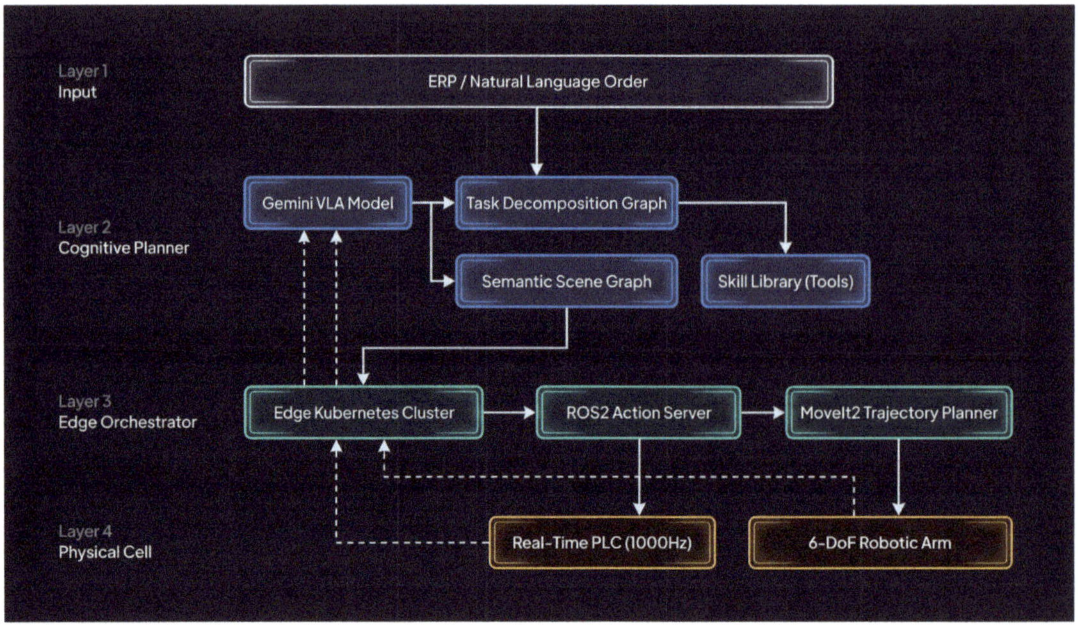

Figure 1.15: Hierarchical architecture of the Robot Cell Task Planner, illustrating the separation of cognitive planning (Agentic Layer) from deterministic execution (Edge/PLC Layer).

The architecture of an Agentic Task Planner is strictly hierarchical, designed to respect the severe latency and safety constraints of Operational Technology (OT) environments. It is divided into three distinct layers:

1. The Cognitive Planner (The Agentic Layer): This layer operates asynchronously, typically in the cloud or on a heavy edge server (e.g., enterprise distributed cloud). It receives high-level instructions, such as "Assemble the planetary gear set, ensuring all components are lubricated." It ingests multimodal context, including CAD models, standard operating procedures (SOPs), and real-time visual feeds from the cell. Its primary output is a Directed Acyclic Graph (DAG) of parameterized skills (e.g., `pick`, `place`, `lubricate`). Because LLM inference can take anywhere from 500 milliseconds to several seconds, this layer is strictly kept out of the robot's real-time control loop.

2. The Edge Orchestrator (The Middleware Layer): Operating on local, air-gapped Kubernetes clusters, this layer acts as the translation engine. It receives the semantic skill requests from the Cognitive Planner and translates them into deterministic commands. In modern robotics, this is almost universally handled by the Robot Operating System (ROS2). The Edge Orchestrator utilizes motion planning libraries like MoveIt2 to calculate the Inverse Kinematics (IK)—the complex mathematics required to determine exactly how the robot's six joints must move to place the end-effector at the target pose without colliding with the environment.

3. The Real-Time Controller (The Physical Layer): This is the domain of the PLC and the robot's proprietary servo drives. Operating at frequencies of 1000Hz (1 millisecond latency), this layer executes the trajectories generated by the Edge Orchestrator. It enforces hard safety interlocks, torque limits, and emergency stops. If the Cognitive Planner hallucinates a command to move the arm through a solid steel table, the Real-Time Controller will reject the trajectory based on its mathematical collision model, ensuring physical safety is never compromised by probabilistic AI.

1.25.1 Grounding the Agent: The Semantic Scene Graph

For an LLM to reason about physical tasks, it must first understand the physical world. Raw camera feeds (RGB) and depth point clouds (D) are meaningless to a text-based reasoning engine unless they are translated into a structured format. This translation is achieved through a **Semantic Scene Graph.**

A Semantic Scene Graph is a dynamic data structure that represents the physical workspace. Nodes in the graph represent objects (e.g., "Gear_A", "Shaft_B", "Robot_Gripper"), and edges represent spatial or semantic relationships (e.g., "is_on_top_of", "is_inside", "distance_to"). When a new task begins, the VLA model processes the camera feeds, runs object detection and pose estimation algorithms, and constructs this graph.

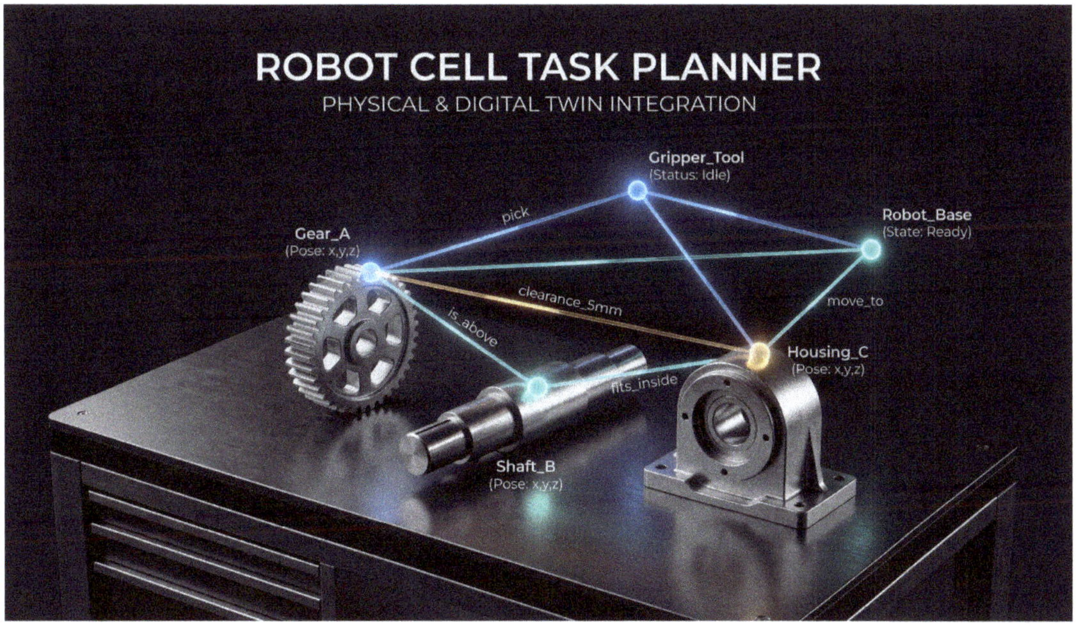

Figure 1.16: A Semantic Scene Graph overlays structured, queryable data onto the unstructured physical workspace, allowing the agent to reason about spatial relationships.

Crucially, the LLM does not need to calculate the exact quaternions or transformation matrices for a 6-Degree-of-Freedom (6-DoF) pose. LLMs are notoriously poor at raw mathematical computation. Instead, the vision system handles the precise pose estimation and stores the numerical coordinates in a state database. The LLM interacts with the *semantic identifiers* of these objects. When the agent decides to `pick("Gear_A")`, the Edge Orchestrator looks up the exact $(x, y, z, roll, pitch, yaw)$ coordinates of "Gear_A" from the state database and passes them to the motion planner. This separation of semantic reasoning from geometric computation is the cornerstone of reliable agentic robotics.

1.26 Task Decomposition and Skill Parameterization

Once the agent understands the state of the world via the Semantic Scene Graph, it must decompose the high-level goal into executable actions. Historically, this was done using Hierarchical Task Networks (HTNs) or Planning Domain Definition Language (PDDL). These symbolic AI approaches required human engineers to manually define every possible state, precondition, and effect in the environment. While mathematically provable, they suffered from combinatorial explosion; introducing a single new object type could require rewriting the entire domain logic.

Agentic planning replaces rigid HTNs with the dynamic reasoning capabilities of Large Language Models. By utilizing the Prompt Chaining and Tool Invocation patterns, the agent can dynamically generate task sequences that adapt to unseen scenarios.

Dimension	Traditional HTN / PDDL	LLM-Based Agentic Planning
State Space Definition	Exhaustive, manually defined	Open-ended, semantically inferred
Handling Unseen Objects	Fails immediately	Generalizes via VLA embeddings
Error Recovery	Requires pre-programmed fallback branches	Autonomous reflection and replanning
Integration Effort	High (Domain experts required)	Medium (Prompt engineering & tool schemas)
Determinism	100% Deterministic	Probabilistic (Requires safety guardrails)

Figure 1.17: Comparison of Traditional Symbolic Planning (HTN/PDDL) versus LLM-Based Agentic Planning in industrial robotics.

1.26.1 The Skill Library: Exposing ROS2 as Tools

To enable the agent to act, the robot's capabilities must be abstracted into a "Skill Library." A skill is a parameterized, atomic action that the robot knows how to execute deterministically. Examples include `move_linear`, `pick_object`, `insert_peg`, or `apply_glue`.

In the context of an Agent Development Kit (ADK) or similar frameworks, these skills are exposed to the LLM as standard tools with strict JSON schemas. The LLM's job is to select the right tool, provide the correct semantic parameters, and sequence them logically.

Consider the following implementation of a Task Planner using Python and a hypothetical ROS2 bridge. The agent is provided with a system prompt that defines its persona, constraints, and access to the skill library.

Listing 1.13: Implementing a Robot Cell Task Planner using Tool Invocation to bridge semantic intent with ROS2 execution.

```
import json
from typing import List, Dict, Any
from pydantic import BaseModel, Field
```

```python
from adk.agents import Agent
from adk.tools import tool

# 1. Define the Tool Schemas (The Skill Library)
class PickParams(BaseModel):
    object_id: str = Field( ... , description="Semantic ID of the object to pick (e.g., 'gear_1').")
    grasp_type: str = Field( ... , description="Type of grasp: 'top_down', 'side', or 'pinch'.")

class PlaceParams(BaseModel):
    object_id: str = Field( ... , description="Semantic ID of the object currently held.")
    target_receptacle_id: str = Field( ... , description="Semantic ID of where to place the object.")
    insertion_force_n: float = Field(default=10.0, description="Downward force limit in Newtons.")

# 2. Implement the Tools (Bridging to Edge Orchestrator)
@tool(name="skill_pick", description="Commands the robot to pick up an object.")
def skill_pick(params: PickParams) -> str:
    """
    In production, this sends a goal to the ROS2 Action Server.
    It looks up the geometric pose of 'object_id' from the Scene Graph database.
    """
    print(f"[ROS2 Bridge] Executing PICK on {params.object_id} via {params.grasp_type} grasp.")
    # Simulated ROS2 synchronous call
    success = ros2_client.execute_action("pick", target=params.object_id, grasp=params.grasp_type)

    if success:
        return f"SUCCESS: Picked up {params.object_id}."
    else:
        return f"FAILURE: Kinematic solver failed to find collision-free path to {params.object_id}."

@tool(name="skill_place", description="Commands the robot to place the held object.")
def skill_place(params: PlaceParams) -> str:
    print(f"[ROS2 Bridge] Executing PLACE of {params.object_id} into {params.target_receptacle_id}.")
    success = ros2_client.execute_action("place", target=params.target_receptacle_id, force=params.insertion_force_n)

    if success:
        return f"SUCCESS: Placed {params.object_id} into {params.target_receptacle_id}."
    else:
        return f"FAILURE: Force limit ({params.insertion_force_n}N) exceeded during insertion. Possible jam."
```

```python
# 3. Initialize the Agentic Planner
system_prompt = """
You are the Cognitive Planner for an industrial robotic cell.
Your goal is to decompose high-level assembly tasks into a sequence of skills.
You have access to the current Semantic Scene Graph.
CRITICAL RULES:
1. You must verify an object exists in the scene graph before interacting with
    it.
2. You must pick an object before you can place it.
3. If a tool returns a FAILURE, you must analyze the error and attempt a
    recovery strategy.
"""

planner_agent = Agent(
    model="vla-advanced-pro",
    system_instruction=system_prompt,
    tools=[skill_pick, skill_place]
)

# 4. Execute a Task
scene_graph_context = "Scene Graph: ['gear_1' (on table), 'housing_A' (in
    fixture), 'shaft_2' (on table)]"
user_order = "Assemble gear_1 into housing_A."

response = planner_agent.run(f"{scene_graph_context}\nOrder: {user_order}")
print(response.text)
```

In this architecture, the LLM acts as a state machine generator. It reasons: "To assemble `gear_1` into `housing_A`, I must first `skill_pick(gear_1, 'top_down')`, and then `skill_place(gear_1, housing_A, 15.0)`." The beauty of this approach is that if a new product variant is introduced—say, `gear_2` which requires a `'side'` grasp—the engineer does not need to rewrite the control logic. They simply update the system prompt or the SOP document in the agent's context window, and the agent adapts its tool parameterization dynamically.

1.27 Closed-Loop Execution and Autonomous Error Recovery

The most significant limitation of traditional robotics is open-loop execution. A PLC assumes the world perfectly matches its internal model. If a part slips in the gripper by 5 millimeters, the robot does not know. It will attempt to insert the part, encounter unexpected resistance, trigger a hardware force-torque alarm, and halt the entire cell. A human operator must then enter the cage, manually jog the robot away, reset the part, and restart the program—a process that can cost thousands of dollars in lost throughput per minute.

Agentic Task Planners solve this by implementing the **Sense-Plan-Act-Reflect** (**SPAR**) lifecycle. The "Reflect" phase is what transforms a blind automaton into an intelligent agent. When a deterministic skill fails at the edge layer, it does not result in a hard halt of the cognitive loop. Instead, the failure is caught by the Edge Orchestrator and returned to the Agentic Planner as a semantic observation.

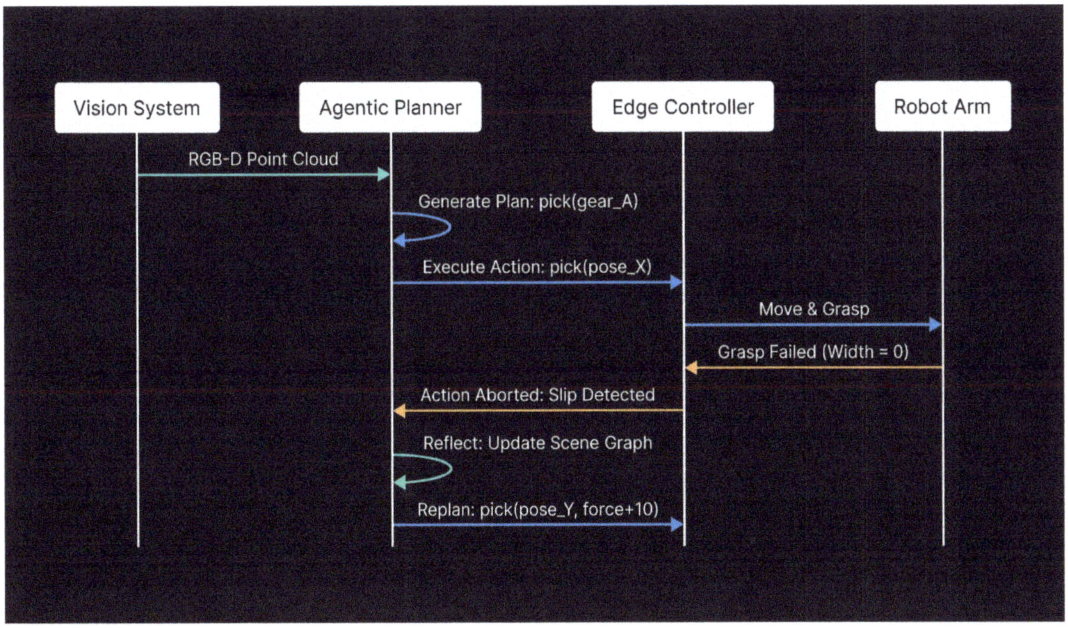

Figure 1.18: Sequence diagram of the SPAR loop. A physical failure (gripper slip) is translated into a semantic observation, allowing the agent to reflect and autonomously generate a recovery plan.

Consider the scenario where the `skill_place` tool returns: `"FAILURE: Force limit (10.0N) exceeded during insertion. Possible jam."`

In a traditional setup, this is a fatal error. In an agentic setup, the LLM receives this string as the output of its tool call. Because the agent is maintaining a conversational state (Memory), it understands the context: it was trying to insert `gear_1`. The agent's system prompt instructs it to attempt recovery. The agent might autonomously decide to: 1. Call a `skill_move_relative(z=50mm)` tool to back the gear out of the housing. 2. Call a `skill_wiggle(amplitude=2mm, frequency=5Hz)` tool to attempt to clear the jam. 3. Call `skill_place` again. 4. If it fails a second time, call a `skill_alert_human(reason="Persistent jam on housing_A")` tool.

This capability—autonomous error recovery—is the holy grail of industrial automation. It drastically increases the Overall Equipment Effectiveness (OEE) by resolving micro-stoppages without human intervention.

1.27.1 Managing Latency and the "Thinking" Pause

A critical engineering consideration when implementing the SPAR loop is latency. LLM inference is slow relative to physical dynamics. If a robot drops a part, the agent might take 2 seconds to process the failure and generate a recovery plan.

During this "thinking" phase, the physical robot must be placed into a safe, stable state. This is managed by the Edge Orchestrator. When an action fails, the ROS2 controller immediately executes a local, deterministic fallback (e.g., "stop all motion and hold current joint positions"). The robot physically pauses. To a human observer, the robot appears to "stop and think." Once the agentic planner returns the new sequence of tool calls, the Edge Orchestrator resumes execution.

This hybrid approach guarantees that the high-latency, non-deterministic reasoning of the LLM never compromises the low-latency, deterministic safety requirements of the physical cell. The agent is the brain, providing the strategy; the edge controller is the brainstem, ensuring immediate survival and reflex actions. By mastering this architectural separation, practitioners can deploy highly adaptive, agentic robotic cells that meet the rigorous demands of modern Industry 4.0 manufacturing floors.

Key Takeaways

- **Decouple Intent from Execution:** Agentic task planners separate high-level cognitive reasoning (handled by LLMs) from low-level kinematic execution (handled by ROS2/PLCs), enabling rapid adaptation to new tasks without manual reprogramming.
- **Semantic Grounding is Essential:** LLMs cannot process raw 6-DoF coordinates effectively. Use Vision-Language-Action models to build Semantic Scene Graphs, allowing the agent to reason using object identifiers rather than raw math.
- **Expose Capabilities via Tool Invocation:** Abstract complex robotic trajectories into parameterized, atomic skills (e.g., `pick`, `place`, `insert`). The agent acts by sequencing these tools and providing semantic parameters.
- **Embrace the SPAR Loop for Error Recovery:** Physical failures are inevitable. Design the system so that deterministic edge failures return as semantic observations to the agent, triggering autonomous reflection and replanning.
- **Enforce Strict Hierarchical Safety:** Never put an LLM in the real-time control loop. The agent dictates the plan asynchronously, while the edge controller enforces millisecond-level safety interlocks, collision avoidance, and emergency stops.

Learning Objectives

- Understand the mathematical and operational limitations of deterministic multi-arm motion planning in dynamic industrial environments.
- Architect a hierarchical agentic system that strictly separates cognitive task planning (semantic reasoning) from kinematic execution (sub-symbolic control).
- Implement dynamic task allocation and temporal synchronization protocols between multiple robotic arms using stateful agent graphs.
- Design self-correcting recovery loops that leverage Vision-Language Models (VLMs) to recover from grasp failures, dropped parts, and dynamic obstacles without human intervention.

The transition from single-arm robotic workcells to multi-arm collaborative environments represents one of the most significant leaps in manufacturing capability—and one of the most profound challenges in control theory. In traditional Industry 4.0 setups, a single robotic arm operates in a highly structured, deterministic environment. The parts arrive at exact coordinates, the arm executes a pre-calculated trajectory, and the operation repeats with sub-millimeter precision. However, when a second, third, or fourth arm is introduced into the same shared workspace, the complexity of the system does not scale linearly; it scales exponentially.

This exponential explosion is rooted in the mathematics of configuration spaces (C-spaces). A standard industrial robotic arm possesses six degrees of freedom (6-DOF). To plan a collision-free path for a single arm, a motion planner must search a 6-dimensional space. When two arms share a workspace, their combined configuration space becomes 12-dimensional. If they are manipulating a shared object, closed-chain kinematic constraints are introduced, further complicating the manifold. Traditional deterministic planners—such as Rapidly-exploring Random Trees (RRT*) or Probabilistic Roadmaps (PRM) operating within frameworks like ROS MoveIt—struggle to compute these high-dimensional trajectories in real-time. To bypass this computational bottleneck, industrial engineers historically resorted to rigid spatial zoning (Arm A never enters Arm B's half of the table) or strict time-division multiplexing (Arm A moves, then stops; Arm B moves, then stops). These workarounds severely limit throughput and entirely preclude true collaboration, such as dynamic hand-offs or dual-arm manipulation of flexible materials.

The advent of Agentic AI introduces a paradigm shift: moving from monolithic, deterministic trajectory calculation to hierarchical, semantic orchestration. A Multi-Arm Coordination Agent does not calculate joint angles. Instead, it acts as the cognitive "foreman" of the workcell. It perceives the environment semantically, reasons about the goals, decomposes the overarching objective into discrete sub-tasks, and delegates these sub-tasks to lower-level, single-arm controllers. By elevating the coordination problem from the sub-symbolic level (geometry and kinematics) to the symbolic level (logic and semantics), agentic architectures bypass the curse of dimensionality that plagues traditional multi-arm systems.

In this chapter, we will dissect the architecture of the Multi-Arm Coordination Agent. We will explore how Large Language Models (LLMs) and Vision-Language Models (VLMs) can be grounded in physical reality using 3D Scene Graphs, enabling them to reason about spatial relationships. We will examine the protocols required for temporal synchronization and dynamic collision avoidance, ensuring that multiple agents can negotiate shared workspace safely. Finally, we will implement a production-ready coordination graph using LangGraph, demonstrating how to bridge the gap between high-level agentic reasoning and low-level ROS2 execution.

1.28 Architectural Foundations of Agentic Multi-Arm Systems

To build a robust Multi-Arm Coordination Agent, we must first establish a strict architectural boundary between cognitive reasoning and kinematic execution. Mixing these two domains is a common anti-pattern that leads to brittle systems. An LLM is exceptionally capable of reasoning that "Arm A should pick up the chassis because it is closer, while Arm B prepares the screws," but it is fundamentally incapable of calculating the inverse kinematics required to move joint 3 by 14.2 degrees to avoid a singularity.

Therefore, the architecture must be strictly hierarchical. At the apex sits the **Meta-Orchestrator Agent**. This agent maintains the global state of the workcell, understands the final assembly goal, and possesses a semantic map of the environment. Below the Meta-Orchestrator are the **Arm Sub-Agents**. Each physical robotic arm is represented by a dedicated sub-agent. These sub-agents are responsible for managing the lifecycle of their specific hardware, including tool changes, payload limits, and local error states. Finally, at the base of the hierarchy are the

Deterministic Motion Planners and **Real-Time Controllers** (e.g., ROS2, OMPL, and hardware PID loops).

This separation of concerns is enforced through well-defined APIs and Tool Calling. The Meta-Orchestrator does not command joint velocities; it invokes tools. A typical tool schema exposed to the Meta-Orchestrator might look like `request_grasp(arm_id="arm_1", target_-object="bracket_A")`. When the LLM decides to invoke this tool, the request is passed down to the Arm 1 Sub-Agent. The Sub-Agent then translates this semantic request into a geometric query: it looks up the current pose of `bracket_A` in the spatial database, calculates a pre-grasp pose, and hands these coordinates to the deterministic motion planner (e.g., MoveIt Task Constructor).

The motion planner then attempts to find a valid, collision-free trajectory through the 6-DOF space. If it succeeds, the trajectory is sent to the real-time controller for execution. If it fails—perhaps because an obstacle is blocking the kinematic path—the motion planner returns an error code to the Sub-Agent, which in turn bubbles a semantic error back up to the Meta-Orchestrator: `"Execution failed: Kinematic path to bracket_A is obstructed by tool_tray"`.

This feedback loop is the crux of agentic autonomy. In a traditional system, an obstructed path results in a hard fault, requiring a human operator to clear the cell and reset the program. In an agentic system, the Meta-Orchestrator receives the semantic error, updates its internal state, and replans. It might decide to invoke a new tool: `move_object(arm_id="arm_2", target_object="tool_tray", destination="storage_zone")`, effectively using the second arm to clear the obstacle for the first arm.

To make this architecture function in a production environment, we must address the impedance mismatch in operating frequencies across the three control layers, summarized in Table~1.2.

Table 1.2: Operating-frequency impedance mismatch across the three control layers

Dimension	Meta-Orchestrator (LLM)	Arm Sub-Agents	Real-Time Controllers
Operating frequency	0.2–1 Hz	10–50 Hz	500–1000 Hz
Responsibility	Semantic orchestration (limited by inference latency)	State machines, trajectory monitoring	Motor currents, PID loops

Because the LLM cannot react in real-time, the lower levels must guarantee safety. The agentic layer operates under the assumption of "safe failure." If a human walks into the cell, or if Arm A unexpectedly drops a part into the path of Arm B, the real-time collision avoidance systems (often utilizing depth cameras and dynamic bounding volumes at the ROS2 level) will trigger an emergency stop (E-Stop) or a protective pause within milliseconds. The LLM is not responsible for preventing the collision; it is responsible for figuring out what to do *after* the protective pause has occurred. Once the hardware is safe, the agent queries the VLM to understand the new state of the world, formulates a recovery plan, and resumes operations.

1.29 Spatial Reasoning and Dynamic Task Allocation

For a Meta-Orchestrator to assign tasks effectively, it must possess a deep, semantic understanding of the physical workspace. Feeding raw coordinate data (e.g., $x = 1.2, y = -0.5, z = 0.8$) to an LLM is highly inefficient. While frontier models exhibit some spatial reasoning capabilities, they struggle to reliably compute Euclidean distances or detect geometric intersections from raw floating-point numbers. Instead, we must translate the physical world into a data structure that LLMs excel at processing: graphs and text.

The industry standard for this translation is the **3D Scene Graph** (3DSG). A scene graph is a hierarchical network where nodes represent physical entities (robots, tools, parts, obstacles, zones) and edges represent the spatial and semantic relationships between them.

To construct a scene graph dynamically, the workcell relies on a perception pipeline. Overhead RGB-D cameras (like Intel RealSense or Azure Kinect) capture point clouds of the workspace. Object detection models (such as YOLOv9 or Segment Anything) identify the objects, while pose estimation algorithms determine their 6-DOF orientation. This geometric data is then passed through a spatial reasoning heuristic engine that calculates relationships. If the bounding box of `part_A` is entirely contained within the bounding box of `bin_1`, the engine creates an edge: `(part_A) -[INSIDE]-> (bin_1)`. If `arm_2` is physically closer to `part_A` than `arm_1`, the engine creates an edge: `(arm_2) -[CLOSEST_TO]-> (part_A)`.

This scene graph is serialized into JSON or YAML and injected into the Meta-Orchestrator's prompt context. By reading the scene graph, the LLM instantly grasps the topology of the workcell without doing any math.

Consider a scenario where a heavy chassis needs to be moved from a conveyor to an assembly fixture. The Meta-Orchestrator receives the following goal: `Assemble(chassis_1, fixture_A)`. It queries the scene graph and notes that `chassis_1` has a property `weight: 15kg`. It also checks the properties of its available resources: `arm_1 (payload: 10kg)` and `arm_2 (payload: 10kg)`.

A rigid, pre-programmed system would fail here if it was only scripted for a 5kg chassis. The agentic system, however, reasons through the constraints. It recognizes that neither arm can lift the chassis alone. It dynamically formulates a new plan: a dual-arm cooperative lift. It allocates the task to both arms simultaneously, issuing commands to the sub-agents to grasp opposite ends of the chassis.

Task allocation in multi-arm systems can be modeled as a market economy or a centralized dispatch. In a centralized dispatch (the most common pattern for 2-4 arms), the Meta-Orchestrator evaluates a cost function for each potential assignment. The cost function considers distance (which arm is closer?), tool availability (does Arm A already have the suction gripper attached?), and kinematic reachability (can Arm B reach the target without entering a singularity?). The LLM acts as the heuristic evaluator, rapidly pruning the decision tree and selecting the optimal assignment.

For more complex environments (e.g., a factory floor with 10+ mobile manipulators), a market-based bidding system is preferred. The Meta-Orchestrator broadcasts a task: `"Need 1x bracket moved to Station 4"`. The individual Arm Sub-Agents calculate their own local cost to complete the task (factoring in their current trajectory, payload, and distance) and reply with a "bid." The Orchestrator simply selects the lowest bid. This decentralized approach reduces the cognitive load on the central LLM and scales linearly with the number of robots.

1.30 Synchronization, Hand-offs, and Collision Avoidance

The most perilous aspect of multi-arm coordination is the shared workspace. When two robots operate in overlapping physical volumes, the risk of catastrophic collision is high. In deterministic systems, this is managed by computing a massive, joint configuration space and finding a single, globally collision-free path for both arms simultaneously. As discussed, this is computationally prohibitive for real-time dynamic environments.

Agentic systems solve this through **Temporal Logic** and **Dynamic Resource Reservation**. Instead of planning every millimeter of the path together, the agents negotiate for space using semantic locks, much like threads negotiating for memory in a computer operating system.

When the Meta-Orchestrator generates a plan, it outputs a Dependency Graph—a Directed Acyclic Graph (DAG) of sub-tasks. For example, a hand-off operation might look like this: 1. Arm A picks up Part X. 2. Arm A moves to Hand-off Zone. 3. Arm B moves to Hand-off Zone. 4. Arm B grasps Part X. 5. Arm A releases Part X. 6. Arm B moves to Assembly Fixture.

The temporal synchronizer enforces the execution order. Step 4 cannot begin until Steps 2 and 3 report a `SUCCESS` state. This ensures that Arm B does not attempt to grasp empty space before Arm A arrives.

However, what happens if Arm A and Arm B need to cross paths to complete unrelated tasks? This is where Dynamic Resource Reservation comes into play. The shared workspace is divided into semantic zones (e.g., `Zone_Center`, `Zone_Left`, `Zone_Right`). Before an Arm Sub-Agent commands its motion planner to generate a trajectory, it must request a "lock" on the zones it intends to traverse.

If Arm A holds the lock for `Zone_Center`, and Arm B's motion planner determines it needs to pass through `Zone_Center`, Arm B's Sub-Agent will request the lock and be denied. Arm B will then enter a `WAITING` state. If Arm A takes too long, a timeout occurs, and the Meta-Orchestrator is alerted to a potential deadlock. The Orchestrator can then intervene, commanding Arm A to move to a safe home position, thereby releasing the lock and allowing Arm B to proceed.

For highly complex, tightly coupled tasks—such as two arms holding a flexible wire while a third arm solders it—semantic locking is insufficient. In these specific edge cases, the Meta-Orchestrator dynamically groups the arms into a temporary "Macro-Arm." It instructs the lower-level ROS2 controllers to merge their configuration spaces temporarily and use a deterministic multi-arm planner (like a constrained optimization solver) for that specific sub-task. Once the soldering is complete, the Orchestrator dissolves the Macro-Arm, and the sub-agents return to independent, semantically locked operation. This hybrid approach leverages the strengths of both paradigms: agentic flexibility for macro-navigation and deterministic precision for micro-manipulation.

1.31 Implementation: Building a Dual-Arm Assembly Cell

To demonstrate these concepts, we will implement the core logic of a Meta-Orchestrator using LangGraph. This implementation will manage two robotic arms (`arm_left` and `arm_right`) tasked with assembling a product. We will define a state graph that handles perception, planning, execution, and, crucially, error recovery.

First, we define the state schema. The state must hold the current scene graph, the active plan, and any execution errors reported by the ROS2 layer.

Listing 1.14: Defining the State Schema for the Multi-Arm Cell

```
from typing import TypedDict, List, Dict, Any, Optional
from pydantic import BaseModel, Field

class Task(BaseModel):
    task_id: str
    assigned_arm: str
    action: str
    target_object: str
    dependencies: List[str] = Field(default_factory=list)
    status: str = "PENDING" # PENDING, IN_PROGRESS, COMPLETED, FAILED

class CellState(TypedDict):
    scene_graph: Dict[str, Any]
    goal: str
    plan: List[Task]
    execution_errors: List[str]
    current_step: int
```

Next, we define the nodes of our LangGraph. The `perceive_scene` node simulates querying a VLM or a 3D Scene Graph database to get the current state of the physical world. The `allocate_tasks` node uses an LLM to generate the dependency graph of tasks.

Listing 1.15: Task Allocation Node using an LLM

```
from langchain_core.prompts import ChatPromptTemplate
from langchain_google_genai import ChatGoogleGenerativeAI

llm = ChatGoogleGenerativeAI(model="gemini-1.5-pro", temperature=0.0)

def allocate_tasks(state: CellState) -> CellState:
    """Generates a coordinated multi-arm plan based on the scene graph."""

    prompt = ChatPromptTemplate.from_messages([
        ("system", """You are the Meta-Orchestrator for a dual-arm robotic cell.
        Available arms: 'arm_left', 'arm_right'.
        Analyze the scene graph and the goal. Break the goal into discrete tasks.
        Assign tasks to arms based on proximity and payload.
        Ensure temporal synchronization using the 'dependencies' field.
```

```python
        Output a JSON list of tasks."""),
        ("user", "Goal: {goal}\nScene Graph: {scene_graph}")
    ])

    # In a real implementation, we use structured output to enforce the Task schema
    chain = prompt | llm.with_structured_output(List[Task])

    generated_plan = chain.invoke({
        "goal": state["goal"],
        "scene_graph": state["scene_graph"]
    })

    return {"plan": generated_plan, "current_step": 0, "execution_errors": []}
```

The execution node is where the agent interfaces with the deterministic hardware layer. In a production environment, this node would publish messages to ROS2 topics or invoke ROS2 actions (e.g., `/arm_left/follow_joint_trajectory`). For this example, we simulate the execution and introduce a potential failure to demonstrate the recovery loop.

Listing 1.16: Execution and Recovery Routing

```python
def execute_and_monitor(state: CellState) -> CellState:
    """Simulates sending commands to ROS2 and monitoring execution."""
    plan = state["plan"]
    errors = []

    # Find tasks that are PENDING and have all dependencies COMPLETED
    ready_tasks = [
        t for t in plan
        if t.status == "PENDING" and
        all(pt.status == "COMPLETED" for pt in plan if pt.task_id in t.dependencies)
    ]

    for task in ready_tasks:
        print(f"Dispatching to ROS2: {task.assigned_arm} -> {task.action} {task.target_object}")

        # --- SIMULATED ROS2 EXECUTION ---
        # Imagine arm_right tries to pick up a part, but the IK solver fails
        # because an unexpected obstacle is in the way.
        if task.assigned_arm == "arm_right" and task.action == "grasp":
            error_msg = f"Kinematic failure on {task.assigned_arm}: Path to {task.target_object} obstructed."
            errors.append(error_msg)
            task.status = "FAILED"
        else:
            task.status = "COMPLETED"

    return {"plan": plan, "execution_errors": errors}
```

```python
def route_after_execution(state: CellState) -> str:
    """Conditional edge to determine if we need to replan or continue."""
    if len(state["execution_errors"]) > 0:
        return "recover"

    if all(t.status == "COMPLETED" for t in state["plan"]):
        return "end"

    return "execute"
```

When an error occurs, the graph routes to the `recover` node. This is the defining feature of an agentic system. Instead of halting, the agent updates its scene graph (perhaps triggering the VLM to take a new picture to find the obstacle) and asks the LLM to generate a corrective plan.

Listing 1.17: The Recovery Node

```python
def recover_from_error(state: CellState) -> CellState:
    """Handles execution failures by replanning."""
    errors = state["execution_errors"]
    print(f"Initiating recovery for errors: {errors}")

    # In a real system, we would trigger a new perception cycle here
    # updated_scene_graph = trigger_vlm_perception()

    prompt = ChatPromptTemplate.from_messages([
        ("system", """You are recovering from a robotic execution error.
        Review the failed plan and the error messages.
        Generate a NEW plan to clear the error and achieve the original goal.
        You may need to use the other arm to clear obstacles."""),
        ("user", "Errors: {errors}\nOld Plan: {plan}\nGoal: {goal}")
    ])

    chain = prompt | llm.with_structured_output(List[Task])
    new_plan = chain.invoke({
        "errors": errors,
        "plan": state["plan"],
        "goal": state["goal"]
    })

    return {"plan": new_plan, "execution_errors": []}
```

Finally, we compile the LangGraph. This creates a stateful, cyclic workflow that continuously perceives, plans, executes, and recovers until the assembly goal is met.

Listing 1.18: Compiling the Multi-Arm Coordination Graph

```python
from langgraph.graph import StateGraph, END

workflow = StateGraph(CellState)

workflow.add_node("perceive", perceive_scene)
```

```
workflow.add_node("plan", allocate_tasks)
workflow.add_node("execute", execute_and_monitor)
workflow.add_node("recover", recover_from_error)

workflow.set_entry_point("perceive")
workflow.add_edge("perceive", "plan")
workflow.add_edge("plan", "execute")

workflow.add_conditional_edges(
    "execute",
    route_after_execution,
    {
        "execute": "execute", # Loop back to execute next ready tasks
        "recover": "recover",
        "end": END
    }
)

workflow.add_edge("recover", "execute") # Try executing the new recovery plan

app = workflow.compile()
```

This architecture fundamentally alters the economics of robotic deployment. By shifting the burden of edge-case handling from human programmers (writing thousands of lines of `if/else` exception logic) to an adaptive reasoning engine, manufacturers can deploy multi-arm cells in high-mix, low-volume environments where rigid automation was previously cost-prohibitive. The agent does not need to be pre-programmed for every possible dropped part or misplaced bin; it simply needs to understand the physics of its environment and the semantics of its goal.

Key Takeaways

- **Hierarchical Separation is Mandatory:** Never use LLMs to calculate joint angles or inverse kinematics. LLMs handle semantic orchestration (who does what, and when); deterministic solvers (ROS2/MoveIt) handle the geometry.
- **Grounding via Scene Graphs:** Raw coordinate data is inefficient for LLMs. Translate the physical workspace into a 3D Scene Graph (nodes = objects, edges = spatial relationships) to enable robust spatial reasoning.
- **Temporal Synchronization:** Manage shared workspaces using Dependency Graphs and semantic zone locking. Arms must negotiate for space to prevent deadlocks and collisions.
- **Safe Failure and Recovery:** The real-time hardware layer must guarantee safety (E-Stops, collision prevention). The agentic layer's responsibility is to perceive the state *after* a safe failure and autonomously generate a recovery plan.
- **Dynamic Task Allocation:** Use centralized cost-function evaluation for small cells (2-4 arms) and market-based bidding for large-scale fleets to assign tasks dynamically based on real-time proximity and payload constraints.

1.32 Vision-Based Pick-Place Copilot

Learning Objectives

- Understand the limitations of traditional machine vision (template matching, rigid bounding boxes) in unstructured, high-mix manufacturing environments.
- Architect a Vision-Language-Action (VLA) pipeline that translates semantic scene understanding into deterministic 3D spatial coordinates for robotic execution.
- Implement a LangGraph-based agentic orchestrator capable of reasoning about grasp affordances, part orientation, and sequencing.
- Design robust "Reflect-and-Retry" loops that utilize visual feedback to autonomously recover from dropped parts, occlusions, and kinematic failures.
- Evaluate the trade-offs between Cloud-based Vision-Language Models (VLMs) and Edge-deployed models regarding latency, token budgets, and production reliability.

In the previous section, we explored how a Meta-Orchestrator can dynamically allocate tasks across a multi-arm robotic cell, relying on a simulated 3D Scene Graph to maintain spatial awareness. However, in real-world manufacturing, the physical environment is rarely as pristine or predictable as a simulated state dictionary. Parts arrive in unstructured bins, components shift on conveyor belts, lighting conditions fluctuate, and objects frequently become entangled or occluded.

When an execution error occurs—such as a kinematic failure or a dropped payload—the agentic system must possess the sensory capability to perceive the new state of the world, understand the semantic context of the failure, and generate a corrective plan. This is where the **Vision-Based Pick-Place Copilot** becomes the critical bridge between the cognitive reasoning of Large Language Models (LLMs) and the physical actuation of industrial robotics.

This section delves exhaustively into the architecture, implementation, and production deployment of agentic vision systems. We will move beyond traditional computer vision techniques, exploring how Vision-Language Models (VLMs) like Gemini 3.5 Pro can be integrated with deterministic spatial solvers to create robots that do not just "see" pixels, but actually "understand" affordances, physics, and assembly semantics.

1.32.1 The Semantic Perception Gap in Industrial Robotics

For decades, industrial automation has relied on highly structured environments. Parts are presented to robots in precise fixtures, blister packs, or vibratory bowl feeders. When machine vision is employed, it typically relies on classical techniques: edge detection, template matching, or, more recently, supervised Convolutional Neural Networks (CNNs) like YOLO or Mask R-CNN.

While these traditional computer vision (CV) systems are incredibly fast and accurate in controlled settings, they suffer from a fundamental flaw: they lack *semantic understanding*. A YOLO model trained to detect a "M8 Hex Bolt" will draw a bounding box around every M8 Hex Bolt in a bin. However, it cannot answer the question: *"Which of these bolts is resting on top of the pile, unobstructed, and oriented in a way that my parallel-jaw gripper can grasp it without colliding with the bin walls?"*

This limitation is known as the **Semantic Perception Gap**. Traditional CV provides geometry without context. In high-mix, low-volume (HMLV) manufacturing, or in tasks like bin-picking and kitting, the robot must constantly reason about physics, occlusion, and sequence. Hard-coding the logic to parse overlapping bounding boxes and calculate depth heuristics requires thousands of lines of brittle, edge-case-heavy code.

Enter the Vision-Language-Action (VLA) paradigm. By utilizing multimodal foundation models, we can pass an image of the workspace to a VLM along with a natural language prompt detailing the goal and the physical constraints of the robot. The VLM acts as a semantic filter, identifying not just the object, but its *affordance*—the specific way it can be interacted with. The VLM outputs a 2D coordinate or a segmentation mask of the optimal grasp point, which is then projected into 3D space using depth sensors, providing the deterministic coordinate required by the robot's motion planner.

1.32.2 Architecture of a Vision-Based Pick-Place Copilot

To build a robust Vision-Based Pick-Place Copilot, we must architect a system that strictly separates semantic reasoning from deterministic execution. LLMs and VLMs are probabilistic; they are excellent at deciding *what* to do and *where* to look, but they are terrible at calculating inverse kinematics (IK) or guaranteeing collision-free trajectories.

The architecture consists of four distinct layers:

1. **Sensory Input Layer:** An RGB-D camera (e.g., Intel RealSense, ZED) captures both a high-resolution 2D color image and a registered 3D point cloud (depth map) of the workspace.
2. **Semantic Perception Layer (VLM):** The 2D image is sent to a Vision-Language Model. The agent prompts the VLM with the current goal (e.g., "Find the optimal grasp point for the top-most aluminum bracket"). The VLM returns structured data, typically a 2D pixel coordinate (u, v) representing the grasp point, and a semantic description of the object's orientation.
3. **Spatial Projection Engine:** The 2D pixel coordinate (u, v) is mapped against the registered depth map to extract the Z (depth) value. Using the camera's intrinsic matrix, the (u, v, Z) data is deprojected into a 3D coordinate (X, Y, Z) in the camera's frame, and then transformed into the robot's base coordinate frame using a known extrinsic calibration matrix.
4. **Deterministic Execution Layer:** The 3D coordinate, along with the semantic orientation data, is passed to a deterministic motion planner (like ROS2 MoveIt). MoveIt calculates the joint angles required to reach the target, ensures the path is collision-free, and executes the trajectory.

This architecture ensures that the VLM is only responsible for what it does best: semantic reasoning. The math required to move the physical motors remains strictly within the domain of proven, deterministic robotics software.

1.32.3 Implementing the Perception-to-Action Loop

Let us examine how to implement this architecture using LangChain, Google's Gemini 3.5 Pro (which excels at spatial reasoning and bounding box generation), and a simulated spatial projection engine.

First, we define the structured output we expect from the VLM. We need the VLM to identify the target, provide a 2D bounding box (from which we will extract the center point), and provide a semantic description of the approach angle.

Listing 1.19: Defining the Grasp Affordance Schema

```
from pydantic import BaseModel, Field
from typing import List, Tuple

class BoundingBox(BaseModel):
    ymin: int = Field(description="Minimum Y pixel coordinate (0-1000 scale)")
    xmin: int = Field(description="Minimum X pixel coordinate (0-1000 scale)")
    ymax: int = Field(description="Maximum Y pixel coordinate (0-1000 scale)")
    xmax: int = Field(description="Maximum X pixel coordinate (0-1000 scale)")

class GraspAffordance(BaseModel):
    target_identified: bool = Field(description="True if the target object is visible and graspable.")
    reasoning: str = Field(description="Explanation of why this specific object was chosen (e.g., 'It is on top of the pile').")
    bounding_box: BoundingBox = Field(description="2D bounding box of the optimal grasp area.")
    approach_vector: str = Field(description="Semantic approach vector: 'top_down', 'side_horizontal', or 'angled'.")
    clearance_warning: bool = Field(description="True if the object is close to the bin wall or other obstacles.")
```

Next, we create the LangGraph node responsible for invoking the VLM. We pass the image of the workspace and prompt the model to act as a robotic vision copilot. Note that Gemini 3.5 Pro can return bounding boxes in a normalized $[0, 1000]$ coordinate space, which we must later scale to our actual image resolution.

Listing 1.20: VLM Perception Node using Gemini 3.5 Pro

```
import base64
from langchain_core.messages import HumanMessage
from langchain_google_genai import ChatGoogleGenerativeAI

# Initialize the VLM
vlm = ChatGoogleGenerativeAI(model="gemini-1.5-pro", temperature=0.0)
vlm_with_structured_output = vlm.with_structured_output(GraspAffordance)

def perceive_workspace(state: dict) -> dict:
    """Analyzes the camera image to find the optimal grasp affordance."""

    image_path = state["current_image_path"]
    target_object = state["target_object"]

    with open(image_path, "rb") as image_file:
        encoded_string = base64.b64encode(image_file.read()).decode('utf-8')
```

```python

    prompt_text = f"""
    You are the vision copilot for an industrial robotic arm with a parallel-jaw gripper.
    Your task is to locate the optimal '{target_object}' to pick up.

    Rules:
    1. The object MUST be unobstructed and resting on top of other items.
    2. Provide a bounding box tightly cropped around the center of mass where the gripper should close.
    3. Determine the safest approach vector to avoid collisions.
    4. Return coordinates in a normalized 0-1000 scale.
    """

    message = HumanMessage(
        content=[
            {"type": "text", "text": prompt_text},
            {"type": "image_url", "image_url": {"url": f"data:image/jpeg;base64,{encoded_string}"}}
        ]
    )

    print(f"Querying VLM for target: {target_object}... ")
    affordance = vlm_with_structured_output.invoke([message])

    return {"affordance": affordance}
```

Once the VLM returns the `GraspAffordance`, the system must translate the normalized 2D bounding box into a physical 3D coordinate. This is handled by the Spatial Projection Engine. In a production environment, this involves querying the depth map (e.g., an Open3D point cloud) at the center of the bounding box.

Listing 1.21: Spatial Projection and Execution Node

```python
import numpy as np

def calculate_3d_grasp_pose(state: dict) -> dict:
    """Translates VLM 2D output into a 3D robotic pose."""
    affordance = state["affordance"]

    if not affordance.target_identified:
        return {"execution_status": "FAILED", "error": "Target not found by VLM."}

    # 1. Calculate center of the bounding box (normalized 0-1000)
    bbox = affordance.bounding_box
    center_x_norm = (bbox.xmin + bbox.xmax) / 2
    center_y_norm = (bbox.ymin + bbox.ymax) / 2

    # 2. Scale to actual image resolution (e.g., 1920x1080)
    img_width, img_height = 1920, 1080
    pixel_u = int((center_x_norm / 1000) * img_width)
```

```
    pixel_v = int((center_y_norm / 1000) * img_height)

    # 3. Query the depth map (Simulated here)
    # In reality: depth_z = depth_image[pixel_v, pixel_u]
    depth_z = 0.45  # meters

    # 4. Deproject to 3D Camera Frame using Intrinsic Matrix (Simulated)
    # X_c = (u - c_x) * Z / f_x
    # Y_c = (v - c_y) * Z / f_y
    cam_x = (pixel_u - 960) * depth_z / 1050.0
    cam_y = (pixel_v - 540) * depth_z / 1050.0
    cam_z = depth_z

    # 5. Transform to Robot Base Frame (Simulated Extrinsic Matrix)
    # robot_pose = extrinsic_matrix @ np.array([cam_x, cam_y, cam_z, 1])
    robot_x = cam_x + 0.5  # Offset from base
    robot_y = cam_y
    robot_z = 0.1          # Table height

    # Map semantic approach vector to Euler angles (Roll, Pitch, Yaw)
    orientation = {"roll": 0, "pitch": np.pi, "yaw": 0} # Default top-down
    if affordance.approach_vector == "side_horizontal":
        orientation = {"roll": 0, "pitch": np.pi/2, "yaw": 0}

    grasp_pose = {
        "x": robot_x, "y": robot_y, "z": robot_z,
        **orientation
    }

    print(f"Calculated 3D Grasp Pose: {grasp_pose}")
    print(f"VLM Reasoning: {affordance.reasoning}")

    # --- SIMULATED ROS2 EXECUTION ---
    # success = ros2_client.execute_trajectory(grasp_pose)
    success = True # Simulate successful pick

    return {"execution_status": "COMPLETED" if success else "FAILED", "
    grasp_pose": grasp_pose}
```

This pipeline demonstrates the immense power of the agentic approach. The developer did not write a single line of OpenCV code to threshold colors, find contours, or filter noise. The semantic complexity of "find the unobstructed object" was entirely offloaded to the VLM, while the spatial precision was maintained by the depth sensor and camera intrinsics.

1.32.4 Handling Edge Cases and Self-Correction

The true test of an autonomous manufacturing cell is not how it performs when everything goes right, but how it recovers when things go wrong. In traditional robotics, a dropped part or a failed grasp usually triggers an E-Stop or a red light, requiring a human operator to enter the cell, clear the fault, and reset the program. This downtime destroys Overall Equipment Effectiveness (OEE).

With a Vision-Based Copilot, we can implement a **Reflect-and-Retry Loop**. If the robot's force-torque sensor or gripper feedback indicates that the payload was dropped, the agentic orchestrator does not halt. Instead, it triggers the camera to take a new picture and asks the VLM to diagnose the failure.

Consider a scenario where a cylindrical part slips from the gripper and rolls to the corner of the bin. The agent captures a new image and prompts the VLM:

"The previous grasp failed. The target object was dropped. Analyze the current image. Locate the dropped object. Explain why the previous top-down grasp might have failed, and suggest a new approach vector."

The VLM, analyzing the new image, might respond: *"The object is now resting against the wall of the bin. A top-down grasp will cause the gripper fingers to collide with the bin wall. The object must be approached using a 'side_horizontal' vector."*

The LangGraph orchestrator receives this new `GraspAffordance`, recalculates the 3D pose with the new orientation parameters, and dispatches the recovery trajectory to ROS2. This closed-loop visual reasoning transforms a brittle automated sequence into a resilient, autonomous agent capable of self-healing.

1.32.5 Performance, Latency, and Edge Deployment

While Cloud VLMs like Gemini 3.5 Pro offer unparalleled semantic reasoning, they introduce a critical constraint in manufacturing: **Latency**. A round-trip API call to a cloud VLM, including image base64 encoding, network transmission, and inference, can take anywhere from 800ms to 2500ms.

In a high-speed pick-and-place cell targeting 60 picks per minute (1 pick per second), a 2-second cognitive delay is unacceptable. Therefore, architects must carefully design the deployment topology, balancing the need for deep reasoning against the strict timing requirements of the industrial cycle.

To solve the latency challenge, production systems employ a **Hybrid Edge-Cloud Architecture**:

- **The Fast Path (Edge):** For standard, repetitive picks, the system relies on a smaller, distilled Vision-Language-Action model (like RT-2 or a fine-tuned Gemini Nano) deployed locally on an industrial IPC (e.g., NVIDIA Jetson AGX Orin). These edge models can process images and output grasp coordinates in under 100ms, ensuring the robot maintains its required cycle time. They operate entirely offline, ensuring factory data privacy and immunity to internet outages.
- **The Slow Path (Cloud):** When the edge model encounters a low-confidence scenario—such as a heavily entangled bin, an unrecognized foreign object, or a dropped part requiring complex recovery reasoning—the agentic orchestrator routes the image to the Cloud VLM. The system accepts the 2-second latency penalty because the alternative is a hard fault requiring human intervention, which would cost minutes of downtime.

Furthermore, token budgets must be managed aggressively. High-resolution images consume significant token counts. To optimize costs and latency, the agent should dynamically crop the image. If the robot is only working in Quadrant 1 of the bin, the orchestrator should crop the image to that specific region before sending it to the VLM, reducing the token payload and focusing the model's attention mechanism on the relevant spatial area.

By combining the semantic depth of foundation models with the deterministic precision of spatial projection and the speed of edge inference, the Vision-Based Pick-Place Copilot represents a paradigm shift in industrial automation. It moves the industry away from rigid, hardcoded sequences and toward truly autonomous, adaptable robotic cells capable of handling the chaos of the real physical world.

Key Takeaways

- **Bridge the Semantic Gap:** Traditional CV provides geometry; VLMs provide context. Use VLMs to identify affordances, orientations, and sequencing logic that are too complex for rigid bounding boxes.
- **Strict Separation of Concerns:** Never ask an LLM to calculate inverse kinematics. Use the VLM to generate 2D pixel coordinates, and use a deterministic Spatial Projection Engine (depth maps + camera intrinsics) to translate those into 3D robot poses.
- **Implement Reflect-and-Retry:** Leverage the VLM not just for the initial plan, but for error recovery. When a grasp fails, capture a new image and ask the VLM to diagnose the failure and suggest a new approach vector.
- **Hybrid Edge-Cloud Topology:** Combat latency by deploying fast, distilled VLA models on edge hardware for standard operations, while routing complex edge cases and error recoveries to powerful Cloud VLMs.
- **Optimize Token Payloads:** Reduce latency and API costs by dynamically cropping images to the relevant workspace quadrant before invoking the VLM.

1.33 Changeover & Tooling Scheduler

Learning Objectives

- Understand the principles of Single-Minute Exchange of Die (SMED) and how agentic orchestration eliminates temporal bottlenecks in high-mix manufacturing.
- Architect a multi-agent system that bridges Enterprise Resource Planning (ERP) schedules with physical intralogistics (AMRs) and robotic cell controllers.
- Implement a LangGraph-based state machine for predictive tool pre-fetching and autonomous changeover execution.
- Design closed-loop visual verification systems using VLMs to ensure safe and accurate end-of-arm tooling (EOAT) swaps.
- Develop resilient exception-handling routines for physical hardware failures during automated changeovers.

While vision-based affordance detection solves the spatial challenge of grasping—allowing robots to handle unstructured environments and variable part geometries—modern manufacturing introduces an equally daunting temporal challenge: the continuous reconfiguration of the workcell itself. In the era of High-Mix, Low-Volume (HMLV) production, a robotic cell rarely performs a single task for days on end. Instead, it must seamlessly transition from assembling automotive water pumps in the morning to machining aerospace brackets in the afternoon.

This transition is known as a *changeover*. In traditional manufacturing, changeovers are the enemy of Overall Equipment Effectiveness (OEE). They require a machine to be halted, existing parts to be cleared, new End-of-Arm Tooling (EOAT) to be physically bolted onto the robot, pneumatic and electrical lines to be reconnected, and new software recipes to be loaded and calibrated. This process, governed by the lean manufacturing principle of Single-Minute Exchange of Die (SMED), traditionally relies heavily on human operators and rigid, sequential standard operating procedures (SOPs).

Even with the advent of Automatic Tool Changers (ATCs)—pneumatic or electrical interfaces that allow a robot to drop one gripper and pick up another—the orchestration of the changeover remains a highly brittle, sequential process. The cell controller waits for the current job to finish, signals an operator or an Automated Guided Vehicle (AGV) to bring the new tool magazine, waits for the delivery, executes the swap, and then waits for human verification before resuming. A 15-minute changeover executed four times a shift results in an hour of lost production daily—a catastrophic drain on capital utilization.

Agentic AI fundamentally transforms this paradigm. By elevating the control logic from rigid PLC (Programmable Logic Controller) sequences to a state-aware, predictive multi-agent orchestrator, we can achieve **Zero-Downtime Changeovers**. The Agentic Changeover Scheduler does not merely react to the end of a production run; it anticipates it, pre-fetches physical assets, coordinates multiple independent hardware systems, and uses visual reasoning to verify the safety and accuracy of the physical swap.

1.33.1 Architecture of the Agentic Tooling Scheduler

To orchestrate a physical changeover, an agentic system must bridge the gap between high-level enterprise IT systems (where the schedule lives) and low-level Operational Technology (OT) systems (where the physical execution happens). This requires a hierarchical, multi-agent architecture.

The architecture consists of three primary agentic roles, typically orchestrated via a framework like LangGraph to maintain strict state management across long-running physical processes:

1. **The Schedule Monitor Agent:** This agent continuously polls the Manufacturing Execution System (MES) or ERP (e.g., SAP, Plex, Ignition) via API. It monitors the current production order's completion percentage. By analyzing the Bill of Process (BOP) for the *upcoming* order, it identifies exactly which EOAT, fixtures, and raw materials will be required next.

2. **The Intralogistics Coordinator Agent:** Once the Schedule Monitor identifies a required tool that is not currently in the cell's local magazine, this agent takes over. It interfaces with the facility's Autonomous Mobile Robot (AMR) fleet manager (e.g., MiR Fleet, OTTO Motors). It calculates the travel time from the centralized tool crib to the robotic cell and dispatches an AMR to deliver the required tooling *just-in-time*, before the current job finishes.

3. **The Cell Execution Copilot:** This is the local agent residing on the cell's IPC. It coordinates the actual physical swap. It commands the robot arm (via ROS2 or native vendor APIs like FANUC's PCDK) to move to the tool drop-off position, actuate the pneumatic ATC, pick up the new tool, and run the calibration sequence.

This separation of concerns is critical. The Schedule Monitor operates in the semantic realm of orders and times; the Intralogistics Coordinator operates in the spatial realm of factory floor routing; and the Cell Execution Copilot operates in the deterministic, real-time realm of kinematics and I/O.

1.33.2 Predictive Scheduling and State-Space Planning

The core advantage of the agentic approach is the shift from sequential execution to parallel pre-fetching. In a traditional setup, the trigger to fetch new tooling is the completion of the current job. In an agentic setup, the trigger is a predictive threshold.

Let us examine how this predictive logic is implemented using LangGraph. The state of the graph must encapsulate both the digital state (MES orders) and the physical state (AMR locations, Robot tool status).

Listing 1.22: Defining the Changeover State and MES Polling Node

```python
from typing import TypedDict, Optional, List
from langgraph.graph import StateGraph, END
import requests
import time

class ChangeoverState(TypedDict):
    current_order_id: str
    current_completion_pct: float
    next_order_id: Optional[str]
    required_tool_id: Optional[str]
    tool_location: str # 'crib', 'in_transit', 'at_cell', 'mounted'
    amr_dispatch_id: Optional[str]
    changeover_status: str # 'idle', 'pre_fetching', 'swapping', 'verifying', 'complete'

def monitor_mes_schedule(state: ChangeoverState) -> dict:
    """Polls the MES to check production progress and upcoming requirements."""

    # Simulated MES API Call
    # response = requests.get(f"http://mes.factory.local/api/orders/{state['current_order_id']}")
    # data = response.json()

    # Simulated data for demonstration
    current_pct = state.get("current_completion_pct", 0.0) + 10.0

    updates = {"current_completion_pct": current_pct}

    # If we are nearing completion (e.g., 80%), look ahead to the next order
    if current_pct >= 80.0 and state["changeover_status"] == "idle":
        print(f"[MES Monitor] Order {state['current_order_id']} at {current_pct}%. Looking ahead...")

        # Simulated look-ahead API call
        next_order = "ORD-992"
```

```python
        required_tool = "GRIPPER-PARALLEL-80MM"

        updates["next_order_id"] = next_order
        updates["required_tool_id"] = required_tool
        updates["changeover_status"] = "pre_fetching"
        print(f"[MES Monitor] Next order {next_order} requires {required_tool}. Initiating pre-fetch.")

    return updates

def route_based_on_status(state: ChangeoverState) -> str:
    """Conditional routing based on the changeover phase."""
    status = state["changeover_status"]
    if status == "idle":
        return "monitor_mes" # Loop back and keep monitoring
    elif status == "pre_fetching" and state["tool_location"] == "crib":
        return "dispatch_amr"
    elif status == "pre_fetching" and state["tool_location"] == "at_cell":
        # Wait until current order is 100% complete before swapping
        if state["current_completion_pct"] >= 100.0:
            return "execute_swap"
        else:
            return "monitor_mes"
    elif status == "swapping":
        return "verify_tool"
    elif status == "complete":
        return END
    return "monitor_mes"
```

In this implementation, the agent acts as a continuous background process. Because LLM inference can be expensive and slow, the `monitor_mes_schedule` node is typically implemented as a deterministic Python function rather than an LLM call. The LLM's reasoning capabilities are reserved for complex decision-making, such as resolving scheduling conflicts (e.g., "The required tool is currently undergoing maintenance; query the MES for an alternative compatible tool or reschedule the order queue").

1.33.3 Autonomous Physical Execution and Tool Swapping

Once the AMR arrives at the cell with the new tool magazine, and the current production order reaches 100% completion, the system transitions to the physical execution phase. This is where the Cell Execution Copilot takes command.

The physical swap involves commanding the robot to move to a specific coordinate in space (the tool stand), disengaging the pneumatic locking mechanism of the current tool, moving to the new tool's location, and engaging the lock. This requires strict deterministic execution. We do not ask an LLM to generate joint angles; instead, we provide the LLM with a set of highly constrained, pre-programmed tools (functions) that trigger ROS2 action servers or PLC routines.

Listing 1.23: Tool Invocation for Physical Execution

```python
from langchain_core.tools import tool

@tool
def execute_atc_swap(current_tool_id: str, target_tool_id: str) -> str:
    """
    Executes the physical Automatic Tool Changer (ATC) sequence.
    Requires the robot to be clear of the workspace.
    """
    print(f"[Robot Controller] Initiating ATC swap sequence.")
    print(f"[Robot Controller] Moving to drop-off stand for {current_tool_id}...")
    # ros2_client.send_goal('move_to_pose', dropoff_pose)

    print(f"[Robot Controller] Disengaging pneumatic lock...")
    # plc_client.write_coil('ATC_LOCK', False)

    print(f"[Robot Controller] Moving to pickup stand for {target_tool_id}...")
    # ros2_client.send_goal('move_to_pose', pickup_pose)

    print(f"[Robot Controller] Engaging pneumatic lock...")
    # plc_client.write_coil('ATC_LOCK', True)

    # Simulate successful execution
    return "SWAP_EXECUTED_PENDING_VERIFICATION"

def execute_swap_node(state: ChangeoverState) -> dict:
    """Node that triggers the physical swap tool."""
    result = execute_atc_swap.invoke({
        "current_tool_id": "CURRENT_TOOL", # In reality, fetched from state
        "target_tool_id": state["required_tool_id"]
    })

    return {"changeover_status": "swapping"}
```

The agentic orchestrator acts as the conductor, ensuring that the `execute_atc_swap` tool is only invoked when all preconditions are met (the AMR is docked, the cell is clear, the previous job is finished). This prevents the catastrophic collisions that often occur in traditional automation when a PLC sequence is triggered out of order.

1.33.4 Closed-Loop Verification and Calibration via VLM

The most critical point of failure in an automated changeover is the mating of the tool. If a metal chip, a drop of coolant, or slight mechanical wear prevents the Automatic Tool Changer from seating perfectly, the pneumatic lock may engage partially. If the robot resumes high-speed production with a partially locked tool, the tool can become a dangerous projectile, destroying the workpiece, the cell, or endangering human life.

Traditional systems rely on simple proximity sensors to verify the lock. However, these sensors are binary and easily fooled by debris. Agentic systems introduce a much more robust paradigm: **Vision-Language Model (VLM) Verification**.

After the physical swap is executed, the robot is commanded to present the wrist joint to a high-resolution inspection camera mounted inside the cell. The image is captured and sent to a VLM (such as Gemini 3.5 Pro or an edge-deployed VLA) with a highly specific prompt.

The prompt to the VLM is engineered to act as a rigorous quality inspector:

"You are a safety inspection copilot for an industrial robotic cell. Analyze the provided image of the Automatic Tool Changer (ATC) mating interface. 1. Verify that the master plate and tool plate are flush, with no visible gap greater than 0.5mm. 2. Verify that the pneumatic locking pins (visible in the indicator window) are fully extended to the 'LOCKED' position. 3. Check for any visible debris, cables caught in the interface, or fluid leaks. Respond with a structured JSON containing 'is_locked' (boolean), 'gap_detected' (boolean), and 'reasoning' (string)."

If the VLM returns `is_locked: true` and `gap_detected: false`, the agentic orchestrator updates the state to `complete`, loads the new recipe parameters for Job B, and resumes production. The entire changeover—from fetching the tool to verifying the lock—has occurred with zero human intervention and zero wasted time.

1.33.5 Exception Handling and Dynamic Replanning

The true power of an agentic architecture over a rigid PLC sequence is its ability to handle exceptions gracefully. The physical world is messy. AMRs get blocked by forklifts. Tools get jammed in their magazines. Pneumatic lines lose pressure.

In a traditional PLC architecture, any deviation from the happy path results in a fault state. A red light flashes, an alarm sounds, and the cell sits idle until a human technician arrives, diagnoses the issue, clears the fault, and manually resets the sequence.

An agentic system employs a **Reflect-and-Retry Loop** to attempt autonomous recovery before escalating to a human.

Consider the scenario where the VLM detects a 2mm gap between the master and tool plates, indicating a failed lock. The agent receives this structured failure state. Instead of halting, the agent's LLM core evaluates the failure context and selects a recovery tool.

It might invoke a `retry_mating_sequence` tool, which commands the robot to purge the pneumatic lines (to blow away potential debris), slightly wiggle the wrist joint (to correct minor misalignments), and re-engage the lock. The robot then presents the wrist to the camera again.

If the AMR is delayed in transit, the Schedule Monitor agent can dynamically replan. Instead of stopping the robot when Job A finishes, it might query the MES: *"Are there any other pending orders that utilize the currently mounted tool?"* If yes, it can dynamically pull that order forward in the queue, keeping the robot productive while waiting for the delayed AMR.

This level of dynamic, context-aware orchestration is impossible with traditional ladder logic. By treating the factory floor as a state-space and utilizing LLMs for high-level reasoning and VLMs for physical verification, the Agentic Changeover Scheduler transforms the most painful bottleneck in HMLV manufacturing into a seamless, autonomous background process.

Key Takeaways

- **Parallelize via Prediction:** Agentic schedulers eliminate downtime by monitoring MES completion percentages and pre-fetching physical assets (tools, materials) via AMRs before the current job finishes.
- **Hierarchical Orchestration:** Separate concerns by using different agents for semantic scheduling (ERP/MES), spatial logistics (AMRs), and deterministic execution (Robot/PLC).
- **VLM as a Quality Inspector:** Replace brittle binary sensors with Vision-Language Models to visually verify complex mechanical states, such as the flush mating of an Automatic Tool Changer.
- **Autonomous Exception Handling:** Implement Reflect-and-Retry loops. If a physical swap fails, the agent should attempt programmatic recovery (e.g., air purges, realignment) before escalating to human operators.
- **Dynamic Queue Management:** If physical assets are delayed, the agent can dynamically reorder the production queue to keep the cell running with its currently mounted tooling, maximizing OEE.

Learning Objectives

- Understand the critical distinction between deterministic safety (PLCs) and probabilistic safety (Agentic Sentinels) in industrial environments.
- Implement uncertainty quantification techniques to evaluate Vision-Language Model (VLM) confidence during physical operations.
- Design and deploy a Human-in-the-Loop (HITL) escalation workflow using LangGraph's `interrupt_before` pattern.
- Architect a dual-channel safety system that maintains ISO 10218 compliance while leveraging the contextual reasoning of Large Language Models.
- Transform opaque machine faults into explainable, context-rich dashboards that accelerate human decision-making and reduce Mean Time to Recovery (MTTR).

1.34 The Imperative for Agentic Safety in Industry 4.0

In the previous chapter, we explored the *Changeover & Tooling Scheduler*, demonstrating how an agentic orchestrator can seamlessly coordinate Autonomous Mobile Robots (AMRs), Manufacturing Execution Systems (MES), and robotic cells to execute zero-touch tool swaps. We highlighted the use of Vision-Language Models (VLMs) as intelligent quality inspectors, capable of visually verifying the flush mating of an Automatic Tool Changer (ATC) before resuming high-speed production.

However, introducing probabilistic models—such as LLMs and VLMs—into the control loop of high-mass, high-velocity industrial machinery introduces a profound paradigm shift in risk management. Traditional industrial automation relies on absolute determinism. A proximity sensor is either triggered or it is not. A ladder logic rung evaluates to either true or false. This binary determinism is the bedrock of industrial safety standards, such as ISO 10218 (Safety requirements for industrial robots) and ISO 13849 (Safety of machinery).

Agentic systems, by contrast, are inherently non-deterministic. A VLM analyzing an image of a tool changer might correctly identify a 2mm gap 999 times out of 1,000. But on the 1,000th attempt, a unique combination of lighting glare, oil smudges on the camera lens, and unexpected shadows might cause the model to hallucinate, confidently declaring a faulty connection as "flush and locked." If the agentic orchestrator blindly trusts this output and commands the robot to accelerate to 2,000 mm/s, the heavy end-effector could detach, resulting in catastrophic damage to the cell or fatal injury to nearby personnel.

This brings us to the core architectural challenge of deploying Agentic AI in the physical world: **How do we harness the unparalleled contextual reasoning of AI without compromising the absolute safety guarantees required on the factory floor?**

The answer is the **Human-in-Loop (HITL) Safety Sentinel** pattern. This pattern does not attempt to make the AI perfectly safe; rather, it assumes the AI is fallible and builds a resilient, dual-channel architecture around it. The Sentinel continuously monitors the agent's reasoning, quantifies the model's uncertainty, and dynamically escalates ambiguous or high-risk decisions to a human operator *before* physical execution occurs.

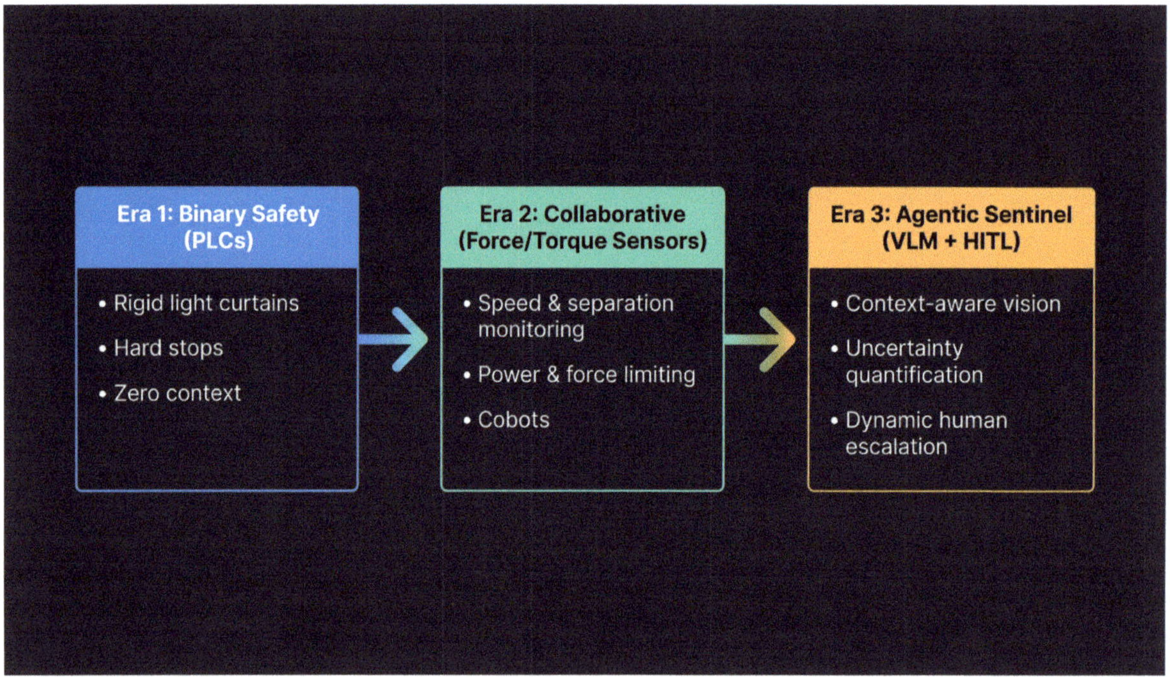

Figure 1.19: The evolution of industrial safety from rigid binary systems to context-aware agentic sentinels.

The Safety Sentinel acts as a cognitive bridge. It translates the rich, unstructured understanding of the physical world generated by the VLM into a structured confidence score. When that confidence drops below a mathematically defined threshold, the Sentinel pauses the autonomous workflow, packages the context (images, reasoning traces, historical data), and presents it to a human expert. The human is no longer a passive monitor waiting for a red light to flash; they become an active, high-level decision node within the agentic graph.

1.35 Dual-Channel Architecture: Deterministic PLCs meet Probabilistic Agents

To satisfy stringent industrial safety regulations (such as achieving a Performance Level 'd' or 'e' under ISO 13849), an agentic system cannot replace the traditional Safety Programmable Logic Controller (Safety PLC). Instead, the architecture must be strictly bifurcated into a **Deterministic Layer** and a **Probabilistic Layer**.

1.35.1 The Deterministic Layer (The Ultimate Authority)

The Deterministic Layer consists of the Safety PLC, hardware interlocks, light curtains, laser scanners, and the robot's internal safety controller. This layer operates on microsecond polling cycles and relies exclusively on binary, hard-wired signals. It has no concept of "context" or "intent." If a human breaks the light curtain, the Safety PLC cuts the contactors, instantly removing motive power from the robot.

The Agentic AI has *zero* ability to override, bypass, or modify the logic of the Deterministic Layer. This is a hard architectural constraint. The Safety PLC is the ultimate, un-hackable authority in the cell.

1.35.2 The Probabilistic Layer (The Contextual Sentinel)

If the Safety PLC is so effective, why do we need an agent? Because deterministic safety is inherently reactive and context-blind. A Safety PLC only stops the robot *after* a dangerous condition has occurred (e.g., a human has already entered the cell, or a tool has already crashed into a fixture). Furthermore, when a PLC halts a cell, it provides minimal context—usually just a cryptic error code like `E-4092: Zone 3 Intrusion`.

The Probabilistic Layer—our Agentic Sentinel—operates above the PLC. It ingests high-bandwidth data: video feeds, audio, MES schedules, and historical maintenance logs. Its goal is to be *predictive* and *contextual*.

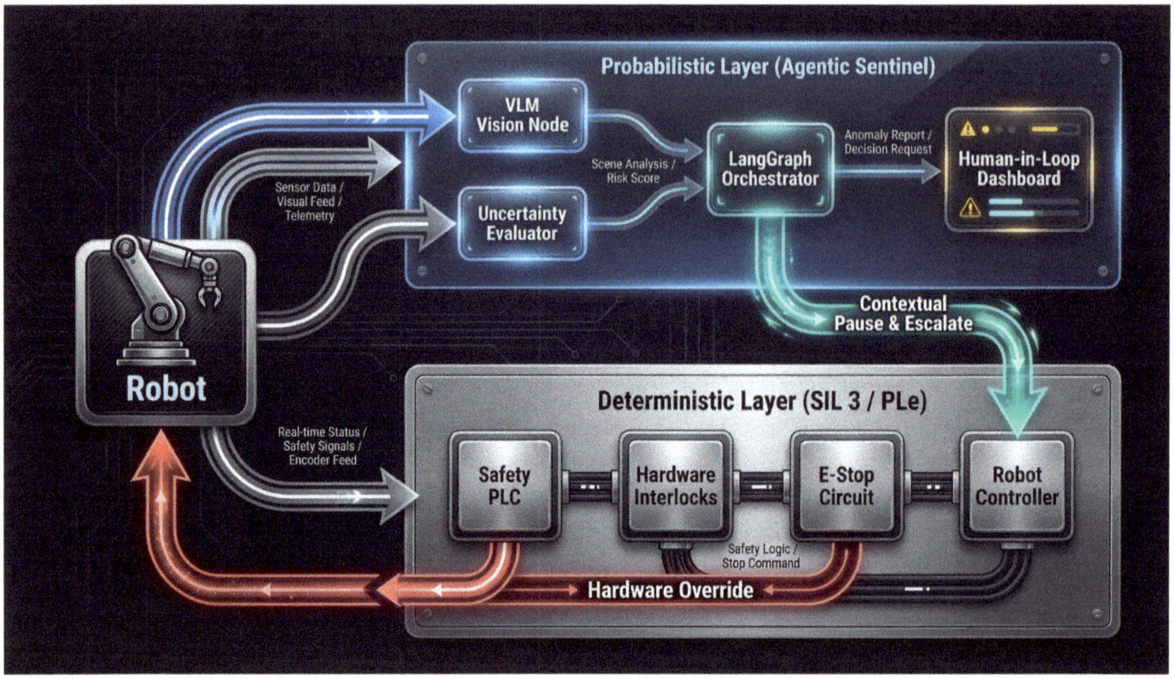

Figure 1.20: Dual-Channel Safety Architecture separating deterministic hardware control from probabilistic agentic reasoning.

While the Sentinel cannot override an E-Stop, it can issue a "Contextual Pause" to the standard robot controller via an API (e.g., ROS2 or OPC UA). For example, if the VLM detects that a tool is misaligned during a changeover, the Sentinel pauses the robot *before* it attempts to move, preventing the crash that would have eventually triggered the Safety PLC.

By catching errors contextually before they become physical collisions, the Sentinel drastically reduces hardware wear, prevents scrapped parts, and transforms a catastrophic fault into a simple human-in-the-loop review task.

1.36 Uncertainty Quantification and Dynamic Thresholding

The core intelligence of the Safety Sentinel lies in its ability to know what it does not know. When we prompt a VLM to verify a tool lock, we cannot simply accept a boolean `True` or `False`. We must quantify the model's epistemic uncertainty (uncertainty due to lack of knowledge) and aleatoric uncertainty (uncertainty due to noisy data, like a blurry camera feed).

In modern agentic frameworks, we achieve this through several techniques:

1. Log-Probability (Logprob) Analysis: When using models like Gemini 3.5 Pro or GPT-5.5, we can request the log probabilities of the generated tokens. If the model outputs `"is_locked": true`, but the logprob for the token `true` is low (e.g., 0.60), it indicates the model is guessing. The Sentinel can enforce a strict threshold (e.g., logprob > 0.95) for critical physical actions.

2. Self-Consistency and Ensemble Prompting: Instead of asking the VLM once, the Sentinel can run the image through the model three times with slightly different temperatures or prompt variations (e.g., "Act as a safety inspector," "Act as a mechanical engineer," "Act as a quality auditor"). If the three outputs disagree, uncertainty is high, and the system escalates to a human.

3. Conformal Prediction: A more advanced statistical method where the model is calibrated on a historical dataset of tool changeovers. Conformal prediction allows the system to output a prediction set with a mathematically guaranteed coverage probability (e.g., "I am 99% confident the state is either 'Locked' or 'Slightly Misaligned'"). If the prediction set contains any unsafe state, the system escalates.

Below is a practical implementation of an Uncertainty Evaluator node within a LangGraph workflow. It uses a combination of structured output and a secondary "Critic" agent to evaluate the primary VLM's confidence.

Listing 1.24: Uncertainty Quantification Node in LangGraph

```python
from typing import TypedDict, Annotated
from langchain_core.messages import HumanMessage, SystemMessage
from langchain_google_genai import ChatGoogleGenerativeAI
from pydantic import BaseModel, Field

# Define the structured output expected from the VLM
class ToolInspectionResult(BaseModel):
    is_locked: bool = Field(description="True if the tool is perfectly flush and locked.")
    gap_detected: bool = Field(description="True if any gap is visible between plates.")
    confidence_score: float = Field(description="Estimated confidence from 0.0 to 1.0.")
    reasoning: str = Field(description="Step-by-step reasoning for the decision.")

class SentinelState(TypedDict):
    image_path: str
    inspection_result: ToolInspectionResult
    requires_human: bool
    escalation_reason: str

# Initialize the VLM (e.g., Gemini 3.5 Pro for multimodal tasks)
vlm = ChatGoogleGenerativeAI(model="gemini-1.5-pro", temperature=0.1)
structured_vlm = vlm.with_structured_output(ToolInspectionResult)

def inspect_tool_node(state: SentinelState) -> dict:
    """Primary VLM inspection of the physical state."""
    # In a real system, we load the base64 image here
    prompt = "Analyze this ATC interface. Is it safely locked?"

    # Simulate VLM invocation
    result = structured_vlm.invoke([HumanMessage(content=prompt)])

    return {"inspection_result": result}

def evaluate_uncertainty_node(state: SentinelState) -> dict:
    """Evaluates the VLM's output against safety thresholds."""
    result = state["inspection_result"]
```

```
    # Rule 1: Hard confidence threshold
    if result.confidence_score < 0.92:
        return {
            "requires_human": True,
            "escalation_reason": f"Low VLM confidence ({result.confidence_score})."
        }

    # Rule 2: Logical contradiction check
    if result.is_locked and result.gap_detected:
        return {
            "requires_human": True,
            "escalation_reason": "Logical contradiction in VLM reasoning."
        }

    # Rule 3: Critic Agent Review (Secondary LLM checks the reasoning)
    critic_prompt = f"""
    Review the following reasoning for a critical safety decision:
    '{result.reasoning}'
    Does this reasoning sound definitive and based on visual evidence,
    or does it contain hedging language (e.g., 'appears to be', 'might be')?
    Respond with 'SAFE' or 'ESCALATE'.
    """
    critic_llm = ChatGoogleGenerativeAI(model="gemini-1.5-flash", temperature=0.0)
    critic_response = critic_llm.invoke([SystemMessage(content=critic_prompt)]).content

    if "ESCALATE" in critic_response:
        return {
            "requires_human": True,
            "escalation_reason": "Critic agent detected hedging in reasoning."
        }

    # If all checks pass, no human needed
    return {"requires_human": False, "escalation_reason": "None"}
```

In this architecture, the agent is actively trying to disprove its own safety hypothesis. Only if the primary VLM is highly confident, logically consistent, and passes a secondary critic's review does the system proceed autonomously. If any check fails, the `requires_human` flag is set to `True`.

1.37 Designing the HITL Escalation Workflow

When the Sentinel determines that human intervention is required, the system must pause gracefully and hand over control. This is where the concept of the **Human-in-the-Loop (HITL) Dashboard** becomes critical.

In traditional automation, an operator responding to a fault walks up to a Human-Machine Interface (HMI) panel, sees a blinking red light and an error code, and must manually investigate the physical cell to understand the context. This leads to high Mean Time to Recovery (MTTR) and operator fatigue.

The Agentic Sentinel transforms this experience. Because the agent has already gathered the context, analyzed the image, and formulated a hypothesis, it can present a highly enriched, explainable dashboard to the operator. The operator is not starting from zero; they are reviewing the agent's homework.

	Traditional HMI Alarm	Agentic HITL Dashboard
Trigger	Binary sensor fault (e.g., proximity switch)	VLM uncertainty threshold exceeded (e.g., logprob < 0.85)
Context Provided	Error Code: E-4092 ATC Fault	Annotated image of tool changer, highlighted 2mm gap, hypothesis: debris on master plate
Human Action Required	Walk to cell, open door, manual inspection, clear fault, reset PLC	Review image on tablet, select 'Purge Air & Retry' or 'Dispatch Maintenance'

Figure 1.21: Comparison of traditional reactive alarms versus context-rich agentic escalation.

1.37.1 Implementing the Interrupt in LangGraph

To implement this technically, we utilize the state graph capabilities of frameworks like LangGraph. LangGraph allows us to define breakpoints in the execution graph using the `interrupt_before` or `interrupt_after` parameters. When the graph execution reaches a node designated as an interrupt, it pauses, persists its state to a database (e.g., PostgreSQL or Redis), and yields control back to the calling application.

The front-end application (the operator's tablet) polls the database, sees the paused graph, and displays the context. The operator makes a decision (e.g., "Approve," "Retry," "Abort"), and the application resumes the graph with the human's input injected into the state.

Listing 1.25: LangGraph Workflow with Human Interrupt

```
from langgraph.graph import StateGraph, END

def route_escalation(state: SentinelState) -> str:
    """Conditional edge to route to human or proceed."""
    if state["requires_human"]:
        return "human_review_node"
    return "execute_physical_action"
```

```python

def human_review_node(state: SentinelState) -> dict:
    """
    This node acts as a placeholder. The graph will be interrupted
    BEFORE executing this node. When resumed, it processes the human input.
    """
    # In a real app, human input is injected into the state upon resume
    human_decision = state.get("human_decision", "abort")

    print(f"[Sentinel] Human reviewed. Decision: {human_decision}")
    return {"human_decision": human_decision}

def execute_physical_action(state: SentinelState) -> dict:
    print("[Robot Controller] Executing safe physical action...")
    return {}

# Build the Graph
workflow = StateGraph(SentinelState)

workflow.add_node("inspect_tool", inspect_tool_node)
workflow.add_node("evaluate_uncertainty", evaluate_uncertainty_node)
workflow.add_node("human_review_node", human_review_node)
workflow.add_node("execute_physical_action", execute_physical_action)

workflow.set_entry_point("inspect_tool")
workflow.add_edge("inspect_tool", "evaluate_uncertainty")

# Conditional routing based on uncertainty
workflow.add_conditional_edges(
    "evaluate_uncertainty",
    route_escalation,
    {
        "human_review_node": "human_review_node",
        "execute_physical_action": "execute_physical_action"
    }
)

# After human review, decide what to do
def route_after_human(state: SentinelState) -> str:
    if state.get("human_decision") == "approve":
        return "execute_physical_action"
    elif state.get("human_decision") == "retry":
        return "inspect_tool" # Loop back
    return END # Abort

workflow.add_conditional_edges(
    "human_review_node",
    route_after_human,
    {
        "execute_physical_action": "execute_physical_action",
        "inspect_tool": "inspect_tool",
        END: END
    }
)
```

```
workflow.add_edge("execute_physical_action", END)

# COMPILE WITH INTERRUPT
# The graph will pause execution right before entering 'human_review_node'
app = workflow.compile(interrupt_before=["human_review_node"])
```

This architecture is incredibly powerful. It treats the human operator not as an external supervisor, but as a specialized, highly reliable "node" within the agentic workflow. The human is invoked exactly like an API call, complete with inputs (the context summary) and expected outputs (the decision).

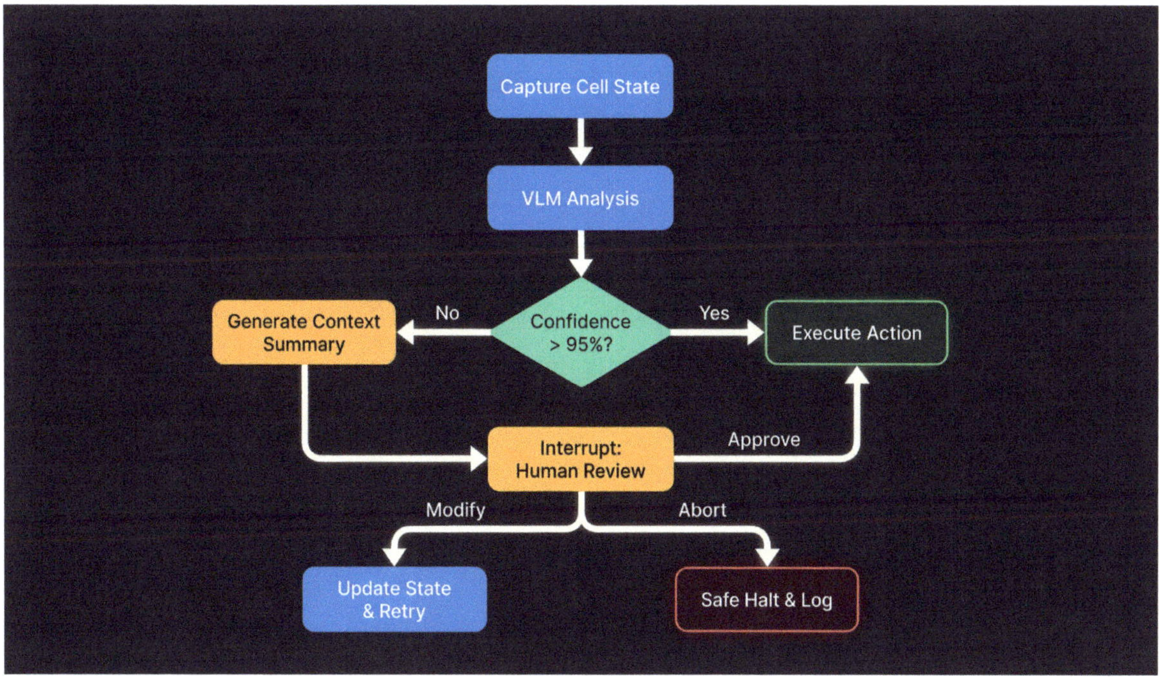

Figure 1.22: State machine representation of the LangGraph workflow, highlighting the human interrupt breakpoint.

1.38 Compliance, Auditability, and Continuous Learning

In heavily regulated industries like manufacturing, aerospace, and automotive, every safety-critical decision must be auditable. When an agentic system makes a decision—or when it defers to a human—the entire context of that decision must be logged immutably.

1.38.1 The Agentic Audit Trail

Traditional PLC logs record binary state changes (e.g., `10:42:01 - Sensor X3 went HIGH`). Agentic audit trails are vastly richer. When the Safety Sentinel executes a workflow, it logs the complete state graph trajectory. This includes:

- The exact image captured by the camera.
- The prompt sent to the VLM.

- The raw JSON output and reasoning trace from the VLM.
- The calculated confidence scores and the output of the Critic agent.
- If escalated, the exact context presented to the human operator.
- The human operator's identity, timestamp, and specific decision (Approve/Retry/Abort).

This rich audit trail is invaluable for root-cause analysis. If a near-miss occurs, safety engineers do not have to guess what the system was "thinking." They can read the exact reasoning trace that led to the action.

1.38.2 Closing the Data Flywheel

The most significant long-term advantage of the Human-in-Loop Safety Sentinel is that it creates a continuous learning loop. Every time the system encounters an edge case it is unsure about, it escalates to a human. The human's decision acts as a high-quality, ground-truth label for that specific edge case.

Over time, this data is aggregated. If the Sentinel frequently escalates due to glare on a specific metallic part, and the human consistently clicks "Approve - Safe," this dataset of images and human approvals can be used to fine-tune a smaller, edge-deployed vision model, or to update the few-shot examples in the VLM's prompt.

As the model learns from the human experts, its confidence on those edge cases increases. The frequency of escalations drops. The system becomes progressively more autonomous, not by removing safety constraints, but by expanding the boundary of what the AI can confidently and safely handle. This is the essence of scalable agentic automation: starting with strict human oversight and gradually earning autonomy through proven, auditable reliability.

Key Takeaways

- **Never Override Deterministic Safety:** Agentic systems must operate as a probabilistic layer *above* the deterministic Safety PLC. The agent can predict and prevent faults, but it cannot bypass hardware interlocks.
- **Quantify Uncertainty:** Do not accept binary answers from LLMs/VLMs for physical actions. Enforce strict confidence thresholds using logprobs, self-consistency checks, or secondary Critic agents.
- **Treat Humans as Graph Nodes:** Use framework features like LangGraph's `interrupt_before` to seamlessly pause execution, persist state, and inject human decisions back into the agentic workflow.
- **Provide Explainable Context:** When escalating to a human, do not just sound an alarm. Provide the VLM's annotated images, reasoning traces, and hypotheses to drastically reduce the operator's cognitive load and MTTR.
- **Leverage Escalations for Learning:** Every human intervention is a labeled data point. Use the audit trail of HITL decisions to fine-tune models and continuously expand the system's safe autonomous envelope.

Learning Objectives

- Understand the physical and network constraints that necessitate edge-native agentic architectures in manufacturing environments.
- Design a distributed orchestration control plane capable of deploying stateful agent graphs and quantized Small Language Models (SLMs) to Industrial PCs (IPCs).
- Implement model quantization and edge-optimized inference engines (e.g., vLLM, TensorRT-LLM) to run reasoning loops on constrained hardware.
- Architect robust state synchronization mechanisms to ensure agent memory persists and reconciles between the edge and the cloud during intermittent network connectivity.
- Deploy a containerized LangGraph agent to an edge node using Kubernetes (K3s) and node selectors.

1.39 The Imperative for Edge-Native Agentic Systems

In the previous chapter, we explored the Human-in-the-Loop (HITL) Safety Sentinel, a system that utilizes Vision-Language Models (VLMs) to evaluate physical cell states, quantify uncertainty, and escalate to human operators when necessary. While the cognitive architecture of that system is robust, its physical deployment introduces a critical vulnerability: the network.

If the VLM and the LangGraph state machine orchestrating the safety checks reside in a centralized cloud region (e.g., Google Cloud `us-central1` or AWS `us-east-1`), every reasoning cycle requires a round-trip over the public internet or a corporate WAN. In the context of enterprise software—generating a marketing email or summarizing a PDF—a latency of 800 milliseconds is imperceptible and entirely acceptable. In the context of Industry 4.0 and robotic control, 800 milliseconds is an eternity.

Consider the physics of a modern manufacturing cell. A standard six-axis industrial robot arm, such as a KUKA KR Quantec or a FANUC R-2000, can move its end-effector at speeds exceeding 2,000 millimeters per second. If an agentic system is responsible for monitoring a dynamic environment and issuing a "stop" or "re-route" command based on visual reasoning, an 800-millisecond cloud latency means the robot will travel 1.6 meters before the command is received. In a collaborative workspace, this is the difference between a safe operation and a catastrophic collision.

Furthermore, factory networks are notoriously hostile environments. They suffer from electromagnetic interference (EMI) generated by heavy machinery, bandwidth saturation during shift changes, and strict firewall policies that often mandate air-gapped operations for critical infrastructure. An agentic system that relies on a continuous, high-bandwidth connection to a cloud API will inevitably fail, leading to unacceptable downtime and compromised safety.

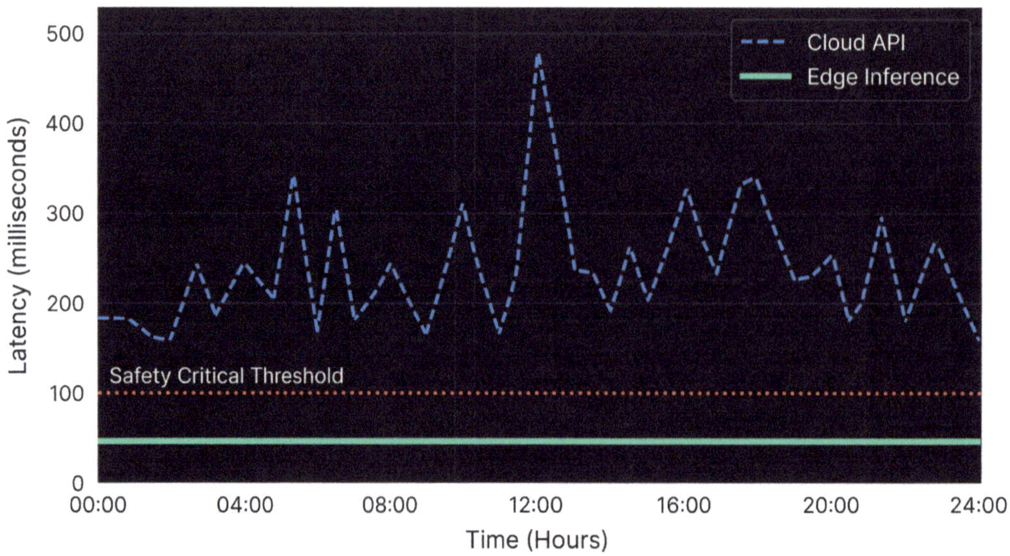

Figure 1.23: Latency comparison: Cloud API calls exhibit high jitter and frequently breach the safety-critical threshold, whereas edge inference provides deterministic, low-latency execution.

To solve this, we must invert the deployment paradigm. Instead of streaming factory data up to the cloud for the agent to reason over, we must push the agent down to the factory floor to reason over the data locally. This is the domain of the **Edge Deployment Orchestrator (EDO)**.

The EDO is a specialized architectural pattern designed to manage the lifecycle, distribution, and execution of agentic workloads on constrained, distributed hardware. It addresses the "Agentic Edge Triad":

1. **Deterministic Latency:** Ensuring that inference and state transitions occur within strict, predictable time bounds (typically < 50ms for near-real-time supervisory control).
2. **Offline Autonomy:** Guaranteeing that the agent can continue to sense, plan, and act even if the uplink to the central cloud is severed.
3. **Data Sovereignty and Privacy:** Processing sensitive proprietary data (e.g., high-resolution images of unreleased product prototypes, precise machining tolerances) entirely on-premises without exposing it to external networks.

Transitioning from cloud-native to edge-native agentic AI is not merely a matter of changing an API endpoint. It requires a fundamental re-engineering of how models are sized, how state is managed, and how updates are orchestrated across a fleet of thousands of disparate industrial machines.

1.40 Architecture of the Edge Deployment Orchestrator

Deploying a traditional machine learning model (like a Random Forest or a simple CNN) to the edge is a well-understood process. You compile the weights, wrap them in a REST API or gRPC service, and deploy a Docker container. Deploying an *agentic system* to the edge is vastly more complex.

An agent is not just a model; it is a stateful, multi-component system. It consists of the core reasoning engine (the LLM/VLM), the cognitive architecture (the LangGraph or CrewAI state machine), the tool execution environment (scripts that interact with PLCs via OPC-UA or Modbus), and the memory systems (local vector databases for semantic search, SQLite for episodic state). The Edge Deployment Orchestrator must manage all of these components synchronously.

1.40.1 The Control Plane vs. The Data Plane

The architecture is strictly divided into a centralized Control Plane (residing in the enterprise cloud, such as Google Cloud or AWS) and a distributed Data/Execution Plane (residing on the factory floor).

The Cloud Control Plane acts as the central nervous system for fleet management. It does not execute the agentic reasoning loops; rather, it is responsible for:

- **Model Registry and Compilation:** Storing the foundational models and fine-tuned adapters (LoRA weights). Crucially, the control plane compiles these models into edge-optimized formats (e.g., GGUF, ONNX, TensorRT engines) tailored to the specific hardware architecture of the target edge node (e.g., ARM64 vs. AMD64, NVIDIA CUDA vs. Intel OpenVINO).
- **Agent Graph Repository:** Version-controlling the Python or TypeScript code that defines the agent's state machine, prompts, and tool schemas.
- **Over-The-Air (OTA) Orchestration:** Using Kubernetes-based fleet management tools (like Google Anthos, Azure Arc, or Rancher Fleet) to push containerized agent workloads to specific factory nodes based on labels and selectors.
- **Global Memory and Analytics:** Aggregating the episodic memory and audit logs synced back from the edge nodes to train better models and monitor fleet-wide performance.

The Factory Edge (Execution Plane) consists of Industrial PCs (IPCs) or specialized edge AI appliances (like the NVIDIA Jetson AGX Orin or Google Distributed Cloud Edge nodes) physically wired into the manufacturing cells. These nodes run:

- **A Lightweight Kubernetes Distribution:** Typically K3s or MicroK8s, which provides container orchestration with a minimal memory footprint.
- **The Local Inference Engine:** High-performance serving frameworks like vLLM, Ollama, or llama.cpp that load the quantized models into local VRAM and expose an OpenAI-compatible API to the local network.
- **The Agent Runtime:** The container executing the LangGraph or CrewAI application. It queries the local inference engine, maintains the state graph, and executes tools.
- **The Local Context Bus:** A high-speed message broker (like MQTT or Redis) that facilitates communication between the agent runtime, the local vector store (e.g., ChromaDB), and the physical machine controllers (PLCs).

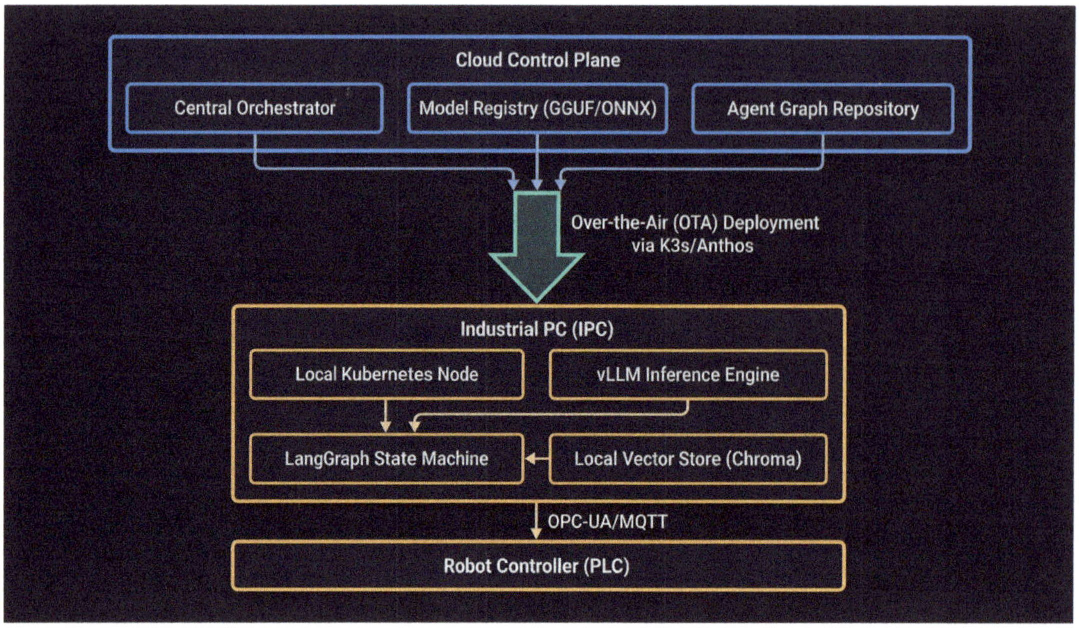

Figure 1.24: The Edge Deployment Orchestrator separates the centralized management of models and graphs from the localized, low-latency execution of the agentic reasoning loop.

1.40.2 State Synchronization and the Split-Brain Problem

One of the most profound challenges in edge agentic architecture is memory management. An agent must maintain context across multiple turns and tasks. In a cloud-native setup, this is easily handled by a managed database like PostgreSQL or Google Cloud Spanner. At the edge, the agent must maintain its state locally to ensure offline autonomy, but this state must eventually be synchronized with the cloud for global analytics, auditability, and cross-factory learning.

This creates a classic distributed systems challenge: the "Split-Brain" problem. If the factory loses internet connectivity, the edge agent continues to operate, generating new episodic memory and updating its state graph. Meanwhile, a human operator in the cloud control plane might attempt to push a configuration update or inject a new goal into the agent's memory. When the network is restored, these two divergent states must be reconciled.

To solve this, the Edge Deployment Orchestrator employs an **Event-Sourced State Sync Engine**. Instead of trying to synchronize the final state of the database, the system records every state transition (every node execution in LangGraph) as an immutable event in a local append-only log (using SQLite or an embedded Kafka instance).

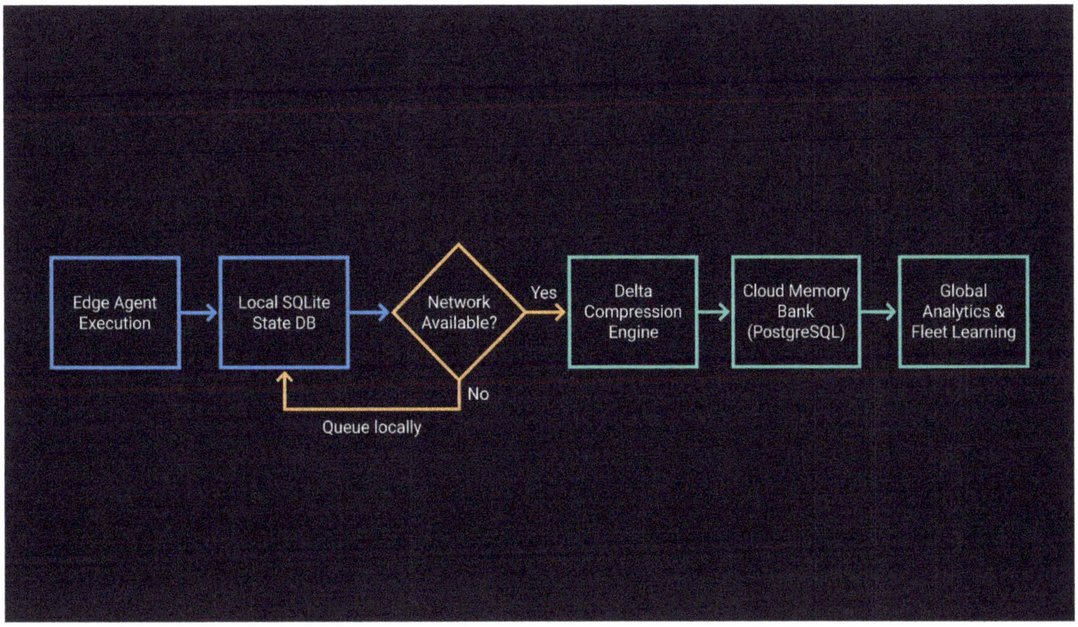

Figure 1.25: Event-sourced state synchronization ensures that edge agents can operate autonomously during network partitions, queuing memory deltas locally until connectivity is restored.

When connectivity is available, a sidecar container reads this local log and streams the deltas to the cloud. Because LangGraph's state management is inherently built on a message-passing architecture (where state updates are applied via reducers), this event-sourced approach maps perfectly to the framework. If conflicts arise, the system utilizes Conflict-free Replicated Data Types (CRDTs) or relies on a strict "Edge Wins" policy for operational data, ensuring that the physical reality of the factory floor always overrides cloud-based directives.

1.41 Model Compression and Edge Runtimes

The most significant physical constraint at the edge is memory bandwidth and VRAM capacity. A frontier model like Gemini 3.5 Pro or GPT-4 requires massive clusters of H100 GPUs to run efficiently. An Industrial PC on a factory floor might only have a single NVIDIA RTX 4060, an integrated Jetson Orin GPU, or even just a standard x86 CPU.

To deploy agentic reasoning to these devices, we must transition from massive, generalized frontier models to **Small Language Models (SLMs)** that have been aggressively compressed and fine-tuned for specific industrial tasks. Models like Google's Gemma 2 (2B and 9B parameters), Meta's Llama 4 (8B), or Microsoft's Phi-3 are the workhorses of the agentic edge.

Feature	Cloud Frontier (e.g., Gemini 1.5 Pro)	Edge SLM (e.g., Gemma 2 9B INT4)
Latency (Time to First Token)	200-800ms (Network Dependent)	< 50ms (Deterministic)
Hardware Requirement	Massive Cloud GPU Clusters	NVIDIA Jetson Orin / 16GB RAM IPC
Offline Capability	Zero (Fails on disconnect)	Full Autonomous Operation
Data Privacy	Data leaves factory floor	100% Air-gapped compliance
Reasoning Depth	Complex multi-step logic, massive context	Focused, task-specific, fine-tuned logic

Figure 1.26: Trade-off matrix comparing Cloud Frontier Models with Edge Small Language Models (SLMs) for industrial agentic applications.

1.41.1 Quantization Techniques

Even an 8-billion parameter model in standard 16-bit floating-point (FP16) precision requires roughly 16GB of VRAM just to load the weights, leaving no room for the KV cache required for context and reasoning. The Edge Deployment Orchestrator relies heavily on **Quantization**—the process of reducing the precision of the model's weights and activations to 8-bit (INT8) or 4-bit (INT4) integers.

Modern quantization techniques are highly sophisticated, ensuring that the model retains its reasoning capabilities despite the massive reduction in precision:

- **AWQ (Activation-aware Weight Quantization):** Unlike naive quantization which treats all weights equally, AWQ observes the model's activations during a calibration phase and preserves the FP16 precision of the top 1% most "salient" weights, while aggressively quantizing the rest to INT4. This results in near-lossless performance for reasoning tasks while reducing the memory footprint by 70%.
- **GGUF (GPT-Generated Unified Format):** A binary format optimized for fast loading and execution on CPUs and Apple Silicon via the `llama.cpp` engine. GGUF allows for mixed-precision quantization (e.g., Q4_K_M), enabling a 9B parameter model to run efficiently on a standard industrial PC without a dedicated GPU, utilizing system RAM and CPU vector instructions (AVX2/AVX-512).

By applying INT4 AWQ quantization, a 9B parameter model's footprint shrinks from 18GB to approximately 5.5GB. This allows the model, the LangGraph runtime, and the local vector database to comfortably coexist on a single edge device with 16GB of unified memory.

1.41.2 High-Performance Edge Inference Engines

Once quantized, the model must be served locally. The orchestrator deploys highly optimized inference engines as sidecar containers alongside the agent runtime.

For GPU-equipped edge nodes, **vLLM** is the industry standard. It utilizes PagedAttention to manage the KV cache efficiently, drastically increasing throughput and reducing latency. For highly constrained or CPU-only environments, **Ollama** (built on `llama.cpp`) provides an incredibly lightweight, self-contained binary that exposes an OpenAI-compatible REST API.

The agent code (written using LangChain or the OpenAI SDK) does not need to know it is running at the edge. It simply points its base URL to `http://localhost:11434/v1` instead of `https://api.openai.com/v1`. This abstraction is crucial: it allows developers to build and test the agentic graph in the cloud using frontier models, and then seamlessly deploy the exact same graph to the edge, swapping only the underlying model endpoint.

1.42 Implementing Edge Orchestration with LangGraph and K3s

To make this concrete, let us examine how an Edge Deployment Orchestrator physically schedules an agent onto a factory node. We will use **K3s**, a lightweight, CNCF-certified Kubernetes distribution designed specifically for IoT and edge computing.

1.42.1 Containerizing the Agent Runtime

First, the agentic application must be packaged. Unlike a standard web app, an edge agent container must include the LangGraph state machine, the tool definitions for local hardware interaction, and the logic to connect to the local inference sidecar.

Listing 1.26: Edge Agent Runtime (edge_agent.py)

```
import os
from langgraph.graph import StateGraph, END
from langchain_openai import ChatOpenAI
from typing import TypedDict

# 1. Connect to the LOCAL inference engine (e.g., Ollama or vLLM running on
    the same IPC)
# The agent is completely air-gapped from the cloud API.
LOCAL_LLM_URL = os.getenv("LOCAL_LLM_URL", "http://localhost:11434/v1")
MODEL_NAME = os.getenv("MODEL_NAME", "gemma2:9b-instruct-q4_K_M")

llm = ChatOpenAI(
    base_url=LOCAL_LLM_URL,
    api_key="edge-local-key", # API key is irrelevant for local inference
    model=MODEL_NAME,
    temperature=0.1 # Low temperature for deterministic industrial control
)

class RobotState(TypedDict):
    sensor_data: str
    anomaly_detected: bool
    action_plan: str
```

```python

def sense_environment(state: RobotState):
    # Tool execution: Read directly from local PLC via OPC-UA
    print("[Edge Agent] Reading local PLC registers... ")
    return {"sensor_data": "Vibration: 4.2mm/s, Temp: 68C"}

def reason_and_plan(state: RobotState):
    prompt = f"Analyze sensor data: {state['sensor_data']}. Is this an anomaly
    ? Plan action."
    response = llm.invoke(prompt)

    # Parse LLM output (simplified for example)
    is_anomaly = "anomaly" in response.content.lower()
    return {"anomaly_detected": is_anomaly, "action_plan": response.content}

def execute_action(state: RobotState):
    if state["anomaly_detected"]:
        print(f"[Edge Agent] Executing local safety halt! Plan: {state['
    action_plan']}")
        # Tool execution: Write to PLC to halt machine
    return state

# Build the LangGraph
workflow = StateGraph(RobotState)
workflow.add_node("sense", sense_environment)
workflow.add_node("reason", reason_and_plan)
workflow.add_node("act", execute_action)

workflow.set_entry_point("sense")
workflow.add_edge("sense", "reason")
workflow.add_edge("reason", "act")
workflow.add_edge("act", END)

app = workflow.compile()

if __name__ == "__main__":
    print(f"Starting Edge Agent connected to {LOCAL_LLM_URL}")
    final_state = app.invoke({"sensor_data": "", "anomaly_detected": False, "
    action_plan": ""})
    print("Cycle complete.")
```

1.42.2 Kubernetes Edge Deployment Manifest

The true power of the Edge Deployment Orchestrator lies in how it distributes this code. Using Kubernetes, we define a `Deployment` that schedules two containers in the same Pod: the Inference Engine (Ollama) and the Agent Runtime.

Crucially, we use `nodeSelector` and `tolerations` to ensure this workload is only deployed to specific IPCs on the factory floor that have the required hardware (e.g., an NVIDIA GPU) and are physically connected to the correct robotic cell.

Listing 1.27: K3s Deployment Manifest for Edge Agent

```yaml
apiVersion: apps/v1
kind: Deployment
metadata:
  name: robotic-cell-agent-alpha
  namespace: factory-floor
spec:
  replicas: 1
  selector:
    matchLabels:
      app: cell-agent
  template:
    metadata:
      labels:
        app: cell-agent
    spec:
      # 1. Node Selector: Ensure this only runs on the specific IPC for Cell Alpha
      nodeSelector:
        factory/location: "assembly-line-1"
        hardware/accelerator: "nvidia-orin"

      containers:
      # Container 1: The Local Inference Engine (Ollama)
      - name: local-llm-engine
        image: ollama/ollama:latest
        ports:
        - containerPort: 11434
        volumeMounts:
        - name: model-cache
          mountPath: /root/.ollama
        # Request GPU resources for the edge node
        resources:
          limits:
            nvidia.com/gpu: 1

      # Container 2: The LangGraph Agent Runtime
      - name: agent-runtime
        image: my-registry.local/edge-agent:v1.2.0
        env:
        # Point the agent to the sidecar inference engine
        - name: LOCAL_LLM_URL
          value: "http://localhost:11434/v1"
        - name: MODEL_NAME
          value: "gemma2:9b-instruct-q4_K_M"

      volumes:
      - name: model-cache
        hostPath:
          path: /mnt/nvme/models # Load quantized models from local fast storage
```

When the central orchestrator applies this manifest, the K3s agent running on the factory IPC pulls the container images, mounts the locally cached GGUF/AWQ models from its NVMe drive, and spins up the Pod. The LangGraph agent immediately begins its sense-plan-act loop, communicating with the local Ollama instance with sub-millisecond network latency.

If the factory's fiber connection to the cloud is severed by a backhoe outside the building, the Kubernetes control plane in the cloud will report the node as `NotReady`. However, because the Pod is already running locally, the Kubelet on the IPC will keep the containers alive. The agent continues to reason, the robot continues to operate safely, and the state deltas are queued in the local SQLite database, waiting patiently for the network to return. This is the ultimate realization of resilient, agentic automation.

Key Takeaways

- **Latency Dictates Architecture:** In industrial robotics, cloud latency is a safety hazard. Agentic reasoning loops must be pushed to the edge to achieve the $<$ 50ms deterministic response times required for physical control.
- **Decouple Control from Execution:** Use a centralized cloud control plane (like Google Anthos or K3s) to manage model registries and graph versions, but execute the actual LangGraph/CrewAI runtime locally on the factory IPC.
- **Embrace Quantization:** Frontier models cannot run on edge hardware. Utilize AWQ or GGUF quantization to compress 8B-9B parameter SLMs (like Gemma 2 or Llama 4) down to 4-bit precision, allowing them to run efficiently on constrained devices.
- **Design for Disconnection:** Factory networks fail. Implement event-sourced state synchronization where the agent writes its memory to a local database first, ensuring full offline autonomy, and syncs deltas to the cloud only when connectivity permits.
- **Abstract the Endpoint:** Write agent code using standard SDKs (OpenAI format), but point the `base_url` to a local sidecar inference engine (vLLM or Ollama). This allows seamless transition from cloud-based development to edge-based production.

Supply Chain & Logistics (MFG focus)

Learning Objectives

- **Understand the paradigm shift** from deterministic Material Requirements Planning (MRP) and static heuristics to probabilistic, multi-agent reasoning in manufacturing supply chains.
- **Architect an Agentic Supply Chain ecosystem** where specialized agents (Procurement, Logistics, Inventory) collaborate, negotiate, and execute actions autonomously via enterprise tool calling.
- **Design robust Human-in-the-Loop (HITL) guardrails** for high-stakes supply chain decisions, ensuring compliance, safety, and financial control during autonomous execution.
- **Implement multi-modal context grounding** to allow agents to reason over structured ERP data, unstructured supplier communications, and real-time external APIs simultaneously.

The modern manufacturing supply chain is a complex, globally distributed, and highly volatile network. For decades, organizations have relied on Enterprise Resource Planning (ERP) systems, Advanced Planning and Scheduling (APS) software, and deterministic Material Requirements Planning (MRP) algorithms to maintain equilibrium. These systems excel at managing structured data and executing predefined mathematical optimizations, such as calculating safety stock based on historical variance or running linear programs for multi-echelon inventory optimization (MEIO). However, they share a critical vulnerability: they are fundamentally reactive and rigid. When a cargo ship is delayed by a geopolitical event, a tier-two supplier sends an unstructured email warning of a raw material shortage, or a sudden spike in social media sentiment alters demand forecasts, traditional systems fail to adapt. They generate exception alerts, cascading the cognitive burden onto human planners who must manually synthesize fragmented data across disjointed dashboards, negotiate via email, and manually update parameters in the ERP. This latency between sensing a disruption and executing a corrective action is the primary driver of the bullwhip effect, excess buffer inventory, and expedited freight costs.

Agentic AI represents a fundamental architectural shift in how supply chains operate, moving from passive systems of record to active, autonomous systems of reasoning. An *Agentic Supply Chain* is an ecosystem of goal-seeking Large Language Model (LLM) agents that continuously perceive their environment, reason over both structured and unstructured data, formulate dynamic plans, and execute actions through enterprise APIs. Unlike traditional machine learning models that merely predict a future state (e.g., forecasting demand), agentic systems possess agency. They do not just predict that a stockout will occur; they autonomously query the ERP for current inventory levels, search the web for alternative suppliers, draft and send negotiation emails to secure emergency stock, and update the Transportation Management System (TMS) to route the new shipment—all while adhering to predefined financial constraints and escalating to a human only when confidence thresholds are breached.

The transition to an agentic architecture requires rethinking the supply chain not as a linear sequence of processes, but as a collaborative multi-agent mesh. In a manufacturing context, a single monolithic agent is an anti-pattern. The domain is too broad, the toolsets too varied, and the context windows too constrained to rely on a single "Supply Chain God Agent." Instead, robust architectures utilize specialized, role-based agents. A Procurement Agent is optimized for negotiation, contract analysis, and supplier relationship management. An Inventory Planner Agent is grounded in inventory holding costs, lead times, and warehouse capacity constraints. A Logistics Agent specializes in routing, carrier selection, and carbon footprint optimization. These agents operate under the supervision of a Sales and Operations Planning (S&OP) Orchestrator, which acts as the meta-agent, decomposing high-level business objectives (e.g., "Maximize Q3 margin while maintaining 98% service level") into specific sub-tasks, delegating them to the specialized agents, and arbitrating conflicts (e.g., when the Logistics Agent wants to use expensive air freight to meet the Inventory Agent's required delivery date).

This multi-agent collaboration is facilitated by standardizing the communication protocols and tool interfaces. Agents communicate via structured Agent-to-Agent (A2A) protocols, passing context, constraints, and proposed plans. They interact with the physical and digital world through a robust tool registry. For instance, when the S&OP Orchestrator detects a potential disruption, it does not merely generate text; it invokes a Python tool to query the SAP HANA database via OData, retrieves the Bill of Materials (BOM), identifies the impacted finished goods, and triggers the Procurement Agent. The Procurement Agent, in turn, uses a Retrieval-Augmented Generation (RAG) tool to search the enterprise contract repository for secondary suppliers, and an Email API tool to initiate contact. This seamless blending of semantic reasoning (the LLM) with deterministic execution (the APIs) is the defining characteristic of production-grade agentic systems.

To ground this conceptually, consider the initialization of a multi-agent supply chain orchestrator using a stateful graph framework like LangGraph. The architecture defines the state schema (which holds the current supply chain context, active disruptions, and proposed plans) and routes execution between specialized agents based on the nature of the task. The following code snippet demonstrates the foundational setup for such an orchestrator, establishing the routing logic between a Procurement Agent and a Logistics Agent.

Listing 1.28: Initializing a Multi-Agent Supply Chain Orchestrator using LangGraph

```python
from typing import TypedDict, Annotated, Sequence
from langgraph.graph import StateGraph, END
from langchain_core.messages import BaseMessage
import operator

# Define the state schema for the Supply Chain Orchestrator
class SupplyChainState(TypedDict):
    messages: Annotated[Sequence[BaseMessage], operator.add]
    active_disruption: str
    impacted_skus: list[str]
    proposed_plan: dict
    requires_human_approval: bool

# Define specialized agent nodes
def sop_orchestrator_node(state: SupplyChainState):
    """Analyzes disruption and delegates to specialized agents."""
    # LLM logic to analyze state.messages and determine next steps
    # ...
    return {"messages": [response_message]}

```

```python
def procurement_agent_node(state: SupplyChainState):
    """Handles supplier negotiation and alternative sourcing."""
    # Invokes SAP API tools and Email tools
    # ...
    return {"messages": [procurement_update]}

def logistics_agent_node(state: SupplyChainState):
    """Optimizes routing and carrier selection."""
    # Invokes TMS API tools and CO2 calculation tools
    # ...
    return {"messages": [logistics_update]}

# Define routing logic
def route_task(state: SupplyChainState):
    last_message = state["messages"][-1].content
    if "SOURCE_MATERIAL" in last_message:
        return "procurement"
    elif "ROUTE_FREIGHT" in last_message:
        return "logistics"
    elif "PLAN_COMPLETE" in last_message:
        return "human_approval" if state["requires_human_approval"] else END
    return "orchestrator"

# Build the Agentic Graph
workflow = StateGraph(SupplyChainState)

workflow.add_node("orchestrator", sop_orchestrator_node)
workflow.add_node("procurement", procurement_agent_node)
workflow.add_node("logistics", logistics_agent_node)

workflow.set_entry_point("orchestrator")
workflow.add_conditional_edges("orchestrator", route_task, {
    "procurement": "procurement",
    "logistics": "logistics",
    "human_approval": END,
    END: END
})
workflow.add_edge("procurement", "orchestrator")
workflow.add_edge("logistics", "orchestrator")

sc_app = workflow.compile()
```

This graph-based approach ensures deterministic execution flow while allowing the LLMs within each node to handle probabilistic reasoning. The orchestrator maintains the global state, ensuring that if the Procurement Agent secures materials from a supplier in Vietnam instead of Mexico, the Logistics Agent is immediately aware and can recalculate the ocean freight lead times and carbon emissions accordingly.

The implementation of these systems yields profound operational benefits. By automating the resolution of routine exceptions, supply chain planners are elevated from data-entry clerks to strategic exception handlers. Organizations deploying agentic supply chains report significant reductions in buffer stock, as the system's ability to react to disruptions in near real-time reduces the need for "just-in-case" inventory. Furthermore, the integration of multi-modal data allows the supply chain to become truly predictive. An agent monitoring global news feeds can detect a port strike in Rotterdam, cross-reference the enterprise BOM to identify critical components in transit, and autonomously reroute shipments to Antwerp before human planners are even aware of the event.

However, deploying autonomous agents in a manufacturing supply chain introduces significant governance and safety challenges. Supply chain decisions carry massive financial and operational weight; an hallucinating agent that autonomously orders ten million units of the wrong microchip, or routes hazardous materials through an unapproved carrier, can cause catastrophic damage. Therefore, the architectural patterns discussed in this chapter heavily emphasize Human-in-the-Loop (HITL) checkpoints, deterministic policy gates, and rigorous evaluation frameworks. We will explore how to constrain agent actions using strict tool schemas, how to implement rollback mechanisms for API calls, and how to design confidence scoring systems that automatically escalate ambiguous situations to human supervisors.

What This Chapter Covers

This chapter provides a comprehensive, hands-on guide to architecting and deploying specialized agents across the manufacturing supply chain lifecycle. We will move beyond theoretical concepts and dive into production-ready design patterns, complete with code implementations and architectural blueprints. The subsequent sections will cover:

- **S&OP Orchestrator:** Designing the meta-agent that aligns sales forecasts with operational capacity, arbitrates conflicts between sub-agents, and maintains the global state graph.
- **Procurement Negotiation Agent:** Building an agent capable of analyzing supplier contracts via RAG, monitoring commodity prices, and autonomously drafting negotiation communications to optimize cost and lead times.
- **Inventory Replenishment Planner:** Implementing an agent that dynamically adjusts safety stock parameters by reasoning over historical ERP data, current demand signals, and warehouse capacity constraints.
- **Transportation Mode Selector:** Architecting a logistics agent that evaluates carrier APIs in real-time to select optimal routing based on cost, urgency, and reliability.
- **Supplier Risk & Disruption Monitor:** Creating an intelligence agent that continuously ingests unstructured web data, news feeds, and weather APIs to proactively identify and mitigate tier-N supplier risks.
- **CO2-Aware Routing Optimizer:** Developing a specialized agent that balances financial cost against environmental impact, ensuring compliance with Scope 3 emissions targets through intelligent freight planning.

By mastering these patterns, technical leaders and AI engineers will be equipped to transform rigid, reactive supply chains into resilient, autonomous ecosystems capable of thriving in an increasingly volatile global market.

Key Takeaways

- **Agentic AI moves supply chains from reactive to autonomous:** Unlike traditional ERPs that rely on deterministic rules and human intervention, agents continuously sense, reason, and act via enterprise APIs.
- **Monolithic agents fail in complex domains:** Production supply chains require a multi-agent architecture where specialized roles (Procurement, Logistics, Inventory) collaborate under an S&OP Orchestrator.
- **Multi-modal reasoning is the differentiator:** Agents unlock value by combining structured enterprise data (SAP/Oracle) with unstructured signals (emails, news, contracts) to resolve exceptions contextually.
- **Stateful orchestration is mandatory:** Frameworks like LangGraph provide the necessary deterministic routing and state management to coordinate probabilistic LLM agents safely.
- **Autonomy requires strict governance:** High-stakes supply chain actions necessitate robust Human-in-the-Loop (HITL) gates, policy-as-code constraints, and confidence-based escalation routing.

1.43 S&OP Orchestrator

Learning Objectives

- **Deconstruct the S&OP Orchestrator** as a meta-agent (Supervisor) responsible for continuous, probabilistic alignment of demand and supply.
- **Design conflict arbitration mechanisms** that allow an LLM to balance competing constraints (e.g., service levels vs. working capital) using corporate objective functions.
- **Implement a stateful routing architecture** using LangGraph to manage delegation between the Orchestrator and specialized sub-agents (Procurement, Logistics, Inventory).
- **Integrate multi-modal context grounding** to fuse structured ERP data (SAP IBP/APO) with unstructured market signals for real-time scenario planning.

1.43.1 The Paradigm Shift to Continuous S&OP

Sales and Operations Planning (S&OP) is the central nervous system of a manufacturing enterprise. Historically, it has been defined as a rigid, cyclical process—typically executed on a monthly cadence—designed to align a company's demand forecast (what sales expects to sell) with its supply plan (what manufacturing and procurement can actually deliver). In traditional environments, this alignment is achieved through a series of siloed, sequential steps: demand planners generate statistical forecasts, supply planners run Material Requirements Planning (MRP) heuristics to determine capacity and material needs, and finance teams evaluate the working capital implications. When discrepancies arise—such as a forecasted demand spike that exceeds current manufacturing capacity—the resolution relies on manual consensus meetings. These meetings are notoriously slow, heavily reliant on static spreadsheets, and prone to human bias.

The fundamental flaw in traditional S&OP is latency. By the time a monthly consensus plan is finalized and published to the Enterprise Resource Planning (ERP) system, the underlying assumptions have often already changed. A critical supplier may have experienced a factory fire, a geopolitical event may have spiked the cost of raw materials, or a viral social media trend may have completely altered the demand curve. Traditional systems, such as SAP Integrated Business Planning (IBP) or Oracle ASCP, are highly capable deterministic engines, but they lack the semantic understanding required to ingest unstructured real-world events and autonomously re-balance the plan in real-time.

The introduction of Agentic AI transforms S&OP from a batch process into a continuous, event-driven orchestration layer. The **S&OP Orchestrator** is not a replacement for the ERP's mathematical solvers; rather, it is a cognitive layer that sits above them. It acts as a meta-agent (or Supervisor) that continuously monitors a multi-modal stream of inputs—structured data from the ERP, unstructured data from supplier emails, and external signals from news and weather APIs. When a disruption is detected, the Orchestrator does not wait for the next monthly meeting. It instantly formulates a hypothesis, delegates sub-tasks to specialized agents (e.g., asking the Procurement Agent to check alternative supplier lead times), arbitrates the inevitable conflicts between those agents, and proposes an updated execution plan. This shift from reactive batch processing to proactive, continuous orchestration is the defining characteristic of the Industry 4.0 supply chain.

1.43.2 Meta-Agent Architecture and Conflict Arbitration

In a complex manufacturing environment, a single monolithic LLM cannot manage the entirety of the supply chain. The context window would be overwhelmed by the sheer volume of SKUs, Bills of Materials (BOMs), routing constraints, and financial parameters. Furthermore, different supply chain functions require different toolsets and optimization objectives. Therefore, the S&OP Orchestrator is designed using the *Supervisor Pattern*.

The Orchestrator itself rarely executes direct API calls to external systems like a Transportation Management System (TMS) or a Warehouse Management System (WMS). Instead, its primary tools are the specialized sub-agents it manages. The Orchestrator maintains the global state, understands the overarching corporate objectives, and routes tasks to the appropriate domain experts.

The most critical function of the S&OP Orchestrator is **Conflict Arbitration**. Supply chain management is fundamentally an exercise in managing trade-offs. The Sales department (represented by a Demand Agent) wants infinite inventory to ensure a 100% Service Level Agreement (SLA) and zero stockouts. The Finance department (represented by a Finance Agent) wants zero inventory to minimize working capital and holding costs. Manufacturing wants long, uninterrupted production runs to maximize Overall Equipment Effectiveness (OEE), while Logistics wants to ship in small, frequent batches to minimize warehouse footprint.

When a disruption occurs—for example, a sudden 20% spike in demand for a flagship product in the EMEA region—the Orchestrator must balance these competing constraints. It queries the Supply Agent, which reports that meeting the demand requires expediting raw materials via air freight. It queries the Finance Agent, which calculates that the air freight costs will erode the product margin by 15%.

To resolve this, the Orchestrator is grounded in a *Corporate Objective Function* provided in its system prompt. This is not a mathematical formula, but a semantic set of prioritized rules. For example:

> *"You are the S&OP Orchestrator for Acme Manufacturing. Your primary objective is to maximize overall corporate margin, subject to the constraint that Tier 1 customer SLAs must never drop below 98%. If a disruption threatens a Tier 1 SLA, you are authorized to incur up to $50,000 in expedited freight costs without human approval. For Tier 2 customers, prioritize margin over SLA, and do not authorize air freight unless the resulting margin remains above 22%."*

Using this semantic grounding, the Orchestrator can reason probabilistically. It evaluates the proposals from the sub-agents, applies the corporate objective function, and makes a deterministic routing decision: it approves the air freight for the Tier 1 customer orders, but delays the Tier 2 orders to the next ocean freight cycle, thereby optimizing the global outcome.

1.43.3 Architecting the Orchestrator Graph with LangGraph

To implement this meta-agent architecture reliably in production, we must move beyond simple prompt chaining and utilize stateful, graph-based orchestration. LangGraph is the industry standard for this pattern, as it allows us to define the S&OP process as a cyclic graph where the Orchestrator node acts as the routing engine, and the specialized agents act as worker nodes.

The state schema is the lifeblood of this architecture. It must contain not only the conversation history (messages) but also structured data representing the current supply chain context, the active disruption being managed, and the financial impact of proposed plans.

The following code demonstrates how to construct the S&OP Orchestrator using LangGraph's `create_team_supervisor` pattern (adapted for a custom state graph to show explicit routing logic). We define the state, the worker nodes, and the LLM-driven routing function that allows the Orchestrator to decide which agent to consult next.

Listing 1.29: Implementing the SOP Orchestrator State Graph

```python
from typing import TypedDict, Annotated, Sequence, Literal
from langgraph.graph import StateGraph, END
from langchain_core.messages import BaseMessage, HumanMessage, AIMessage
from langchain_google_vertexai import ChatVertexAI
from pydantic import BaseModel, Field
import operator

# 1. Define the Global S&OP State
class SOPState(TypedDict):
    messages: Annotated[Sequence[BaseMessage], operator.add]
    active_disruption: str
    impacted_skus: list[str]
    current_margin_impact: float
    sla_at_risk: bool
    next_agent: str # The agent the orchestrator decides to call next

# 2. Define the Routing Schema for the Orchestrator LLM
class OrchestratorDecision(BaseModel):
```

```python
    reasoning: str = Field(description="Step-by-step reasoning for the next action based on corporate objectives.")
    next_agent: Literal["ProcurementAgent", "LogisticsAgent", "FinanceAgent", "FINISH"] = Field(
        description="The next specialized agent to route to, or FINISH if a resolution plan is complete."
    )

# 3. Initialize the Frontier Model (Gemini 3.5 Pro for complex reasoning)
llm = ChatVertexAI(model_name="gemini-3.5-pro", temperature=0.1)
orchestrator_llm = llm.with_structured_output(OrchestratorDecision)

# 4. Define the Orchestrator Node
def sop_orchestrator_node(state: SOPState):
    system_prompt = """You are the S&OP Orchestrator. Your goal is to resolve supply chain disruptions.
    Corporate Objective: Maximize margin. Tier 1 SLAs must remain > 98%.
    Review the conversation history and the current state.
    Decide which specialized agent to consult next to gather information or propose a solution.
    If a viable plan has been formed that meets objectives, output FINISH."""

    messages = [SystemMessage(content=system_prompt)] + state["messages"]
    decision = orchestrator_llm.invoke(messages)

    # Append the orchestrator's reasoning to the message history
    new_message = AIMessage(content=f"Orchestrator Reasoning: {decision.reasoning}\nRouting to: {decision.next_agent}")

    return {
        "messages": [new_message],
        "next_agent": decision.next_agent
    }

# 5. Define Worker Nodes (Mocked for brevity - these would invoke their own LLMs and Tools)
def procurement_node(state: SOPState):
    # In reality, this agent calls SAP Ariba APIs, searches contracts, etc.
    response = AIMessage(content="Procurement: Alternative supplier found in Mexico. Lead time 5 days. Cost +12%.")
    return {"messages": [response]}

def logistics_node(state: SOPState):
    # In reality, this agent calls TMS APIs, calculates CO2, etc.
    response = AIMessage(content="Logistics: Air freight available. Cost $15,000. Delivery in 2 days.")
    return {"messages": [response]}

def finance_node(state: SOPState):
    # Calculates margin impact
    response = AIMessage(content="Finance: Air freight reduces margin to 18% (below 22% threshold for Tier 2).")
    return {"messages": [response], "current_margin_impact": -0.04}
```

```python
# 6. Define the Conditional Routing Edge
def route_to_agent(state: SOPState):
    if state["next_agent"] == "FINISH":
        return END
    return state["next_agent"]

# 7. Build and Compile the Graph
workflow = StateGraph(SOPState)

workflow.add_node("Orchestrator", sop_orchestrator_node)
workflow.add_node("ProcurementAgent", procurement_node)
workflow.add_node("LogisticsAgent", logistics_node)
workflow.add_node("FinanceAgent", finance_node)

workflow.set_entry_point("Orchestrator")

# The Orchestrator routes to one of the agents, or ends the process
workflow.add_conditional_edges("Orchestrator", route_to_agent)

# All worker agents route back to the Orchestrator to update the global state
workflow.add_edge("ProcurementAgent", "Orchestrator")
workflow.add_edge("LogisticsAgent", "Orchestrator")
workflow.add_edge("FinanceAgent", "Orchestrator")

sop_app = workflow.compile()
```

In this architecture, the `sop_orchestrator_node` acts as the brain. It uses `with_structured_output` to force the LLM to output a deterministic routing decision (`next_agent`) alongside its probabilistic reasoning. This ensures the graph executes reliably without parsing errors. The Orchestrator continuously loops, gathering context from the Procurement, Logistics, and Finance agents until it has synthesized a plan that satisfies the corporate objective function, at which point it routes to `END` (which, in a production system, typically triggers a human approval workflow).

1.43.4 Multi-Modal Grounding and ERP Integration

An agent is only as intelligent as the context it can access. Traditional S&OP systems fail because they only see structured data: rows and columns in a database representing inventory levels, lead times, and unit costs. They are blind to the unstructured reality of the supply chain: a PDF contract detailing force majeure clauses, an email thread from a supplier warning of a potential strike, or a news article about a port closure.

The Agentic S&OP Orchestrator leverages multi-modal grounding to fuse these disparate data sources into a unified context window. This is typically achieved using a hybrid retrieval architecture.

For structured data, the Orchestrator utilizes Text-to-SQL or GraphQL tools to query the enterprise data warehouse. In the Google Cloud ecosystem, this is often implemented using BigQuery and the Google Cloud Cortex Framework, which provides pre-built data foundations for SAP environments. When the Orchestrator needs to know the current inventory of SKU-123, it invokes a tool that executes a BigQuery SQL statement against the replicated SAP ECC or S/4HANA tables, returning the exact quantity and location.

For unstructured data, the Orchestrator utilizes Vector Search (e.g., Vertex AI Vector Search). Supplier contracts, historical S&OP meeting minutes, and email communications are embedded and stored in a vector database. If a disruption occurs with a specific supplier, the Orchestrator can retrieve the exact penalty clauses from the contract PDF and factor them into its financial arbitration.

Furthermore, the Orchestrator can be equipped with external API tools. By integrating with services like the GDELT Project (Global Database of Events, Language, and Tone) or specialized supply chain risk APIs (e.g., Everstream Analytics), the Orchestrator can sense disruptions before they manifest in the ERP. If a news API reports a typhoon approaching a major manufacturing hub in Taiwan, the Orchestrator can proactively query the ERP for BOMs reliant on components from that region, identify the risk exposure, and trigger the Procurement Agent to secure alternative supply days before the human planners even receive an alert.

1.43.5 Production Deployment and HITL Governance

Deploying an autonomous S&OP Orchestrator into a production manufacturing environment carries significant risk. Supply chain decisions involve millions of dollars in capital allocation, binding legal contracts, and physical movement of goods. An hallucination at the Orchestrator level—such as erroneously ordering a year's worth of perishable inventory—can be catastrophic.

Therefore, the architecture must incorporate rigorous Human-in-the-Loop (HITL) governance and deterministic policy gates. The Orchestrator should not be granted direct, unconstrained write access to the ERP system. Instead, the architecture follows a "Plan and Propose" pattern.

When the Orchestrator reaches the `FINISH` state in its LangGraph execution, it generates a structured JSON payload representing the proposed execution plan (e.g., PO modifications, route changes, inventory transfers). This payload is first passed through a deterministic Policy Engine (written in standard code, not an LLM). The Policy Engine checks hard constraints: Does this PO exceed the $50,000 automated approval limit? Does this route violate hazardous materials regulations?

If the plan passes the deterministic checks, it is routed to a human supervisor via an approval dashboard (often integrated into Microsoft Teams, Slack, or a custom web application). The Orchestrator presents not just the proposed action, but its complete reasoning trace, the alternatives it considered, and the financial impact analysis generated by the Finance Agent. The human planner reviews the synthesized context, clicks "Approve," and only then does the system execute the write operations back to the ERP and TMS via API.

Over time, as the organization builds trust in the system, the deterministic thresholds can be adjusted. A company might start by requiring human approval for all actions, then gradually allow the Orchestrator to autonomously execute inventory transfers under $5,000, then POs under $10,000, progressively scaling the level of autonomy while maintaining strict financial guardrails. This phased approach ensures safety while unlocking the massive speed and efficiency gains of Agentic AI in the supply chain.

Key Takeaways

- **Continuous over Batch:** The S&OP Orchestrator replaces static, monthly consensus meetings with continuous, event-driven scenario planning, drastically reducing response latency to supply chain disruptions.
- **Supervisor Pattern is Essential:** A single LLM cannot manage the entire supply chain. The Orchestrator acts as a meta-agent, delegating tasks to specialized domain agents (Procurement, Logistics) and maintaining global state.
- **Semantic Arbitration:** The Orchestrator resolves conflicts between competing supply chain metrics (e.g., SLA vs. Working Capital) by reasoning over a semantically defined Corporate Objective Function.
- **Multi-Modal Context is the Differentiator:** True agentic value comes from fusing structured ERP data (via SQL/APIs) with unstructured signals (contracts, news, emails) to form a complete operational picture.
- **Govern via "Plan and Propose":** High-stakes S&OP actions require deterministic policy gates and Human-in-the-Loop (HITL) approval workflows before executing write operations to enterprise systems.

1.44 Procurement Negotiation Agent

Learning Objectives

- Understand the architectural transition from static, rules-based procurement to dynamic, multi-variable agentic negotiation.
- Design a stateful negotiation loop using LangGraph that calculates Best Alternative to a Negotiated Agreement (BATNA) in real-time.
- Implement multi-modal context fusion to evaluate supplier proposals against historical contracts, ERP data, and external market signals.
- Establish deterministic policy guardrails and Human-in-the-Loop (HITL) escalation paths for high-stakes financial commitments.

While the S&OP Orchestrator discussed in the previous section excels at identifying supply chain disruptions and proposing high-level mitigation strategies, the actual execution of those strategies often requires interacting with external entities. When the Orchestrator determines that a critical component shortage requires sourcing from an alternative supplier, the process transitions from internal planning to external execution. This is the domain of the Procurement Negotiation Agent.

Historically, automated procurement has been limited to catalog buying or rigid Request for Proposal (RFP) systems. These systems are highly deterministic but lack the cognitive flexibility to negotiate. They can compare unit prices on a spreadsheet, but they cannot parse a supplier's email explaining that a 5% price premium is required for expedited shipping, nor can they counter-offer by adjusting payment terms to offset the cost. The Procurement Negotiation Agent bridges this gap, utilizing Large Language Models (LLMs) to conduct nuanced, multi-turn negotiations via email, EDI, or supplier portals, optimizing for a complex matrix of business objectives.

1.44.1 The Anatomy of an Autonomous Negotiator

A production-grade Procurement Negotiation Agent is not a single prompt; it is a composite cognitive architecture designed to mimic the strategic reasoning of a human buyer. To achieve this, the agent must maintain state across days or weeks of back-and-forth communication, continuously updating its internal model of the negotiation.

The architecture consists of four primary sub-systems: the Context Engine, the Strategy Planner, the Communication Interface, and the Evaluation Critic.

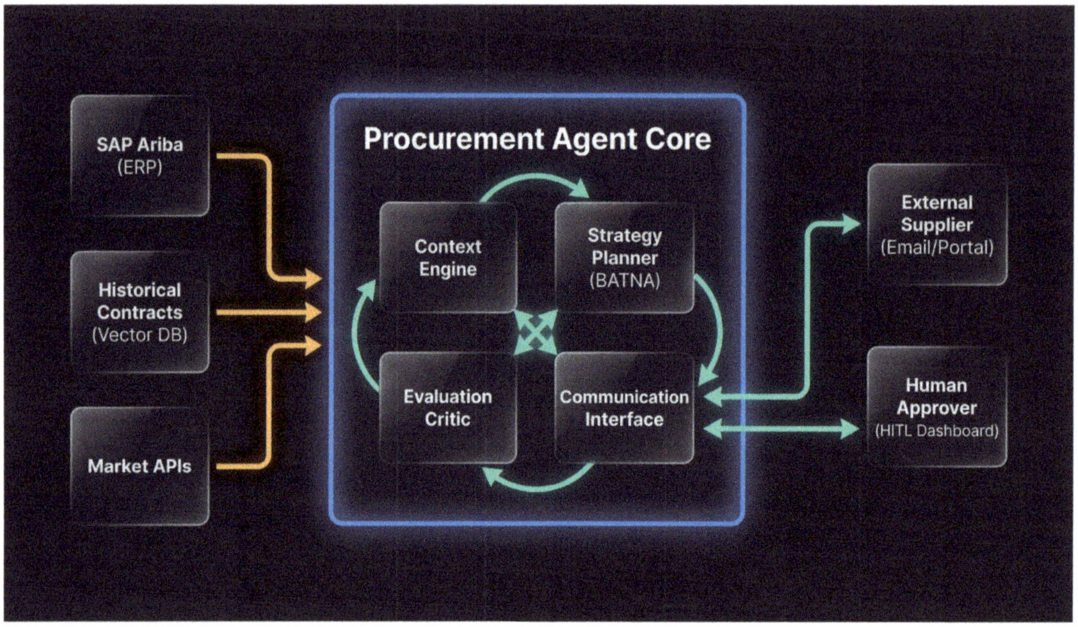

Figure 1.27: Core architecture of the Procurement Negotiation Agent, illustrating the separation of context retrieval, strategic planning, and external communication.

The **Context Engine** is responsible for grounding the agent in reality. Before drafting a single message, it queries the enterprise ERP (e.g., SAP Ariba, Coupa) to retrieve the exact specifications of the required material, the approved budget, and the required delivery date. Simultaneously, it uses Vector Search to retrieve historical contracts with the specific supplier, identifying past concessions, standard payment terms (e.g., Net 60), and historical pricing volatility.

The **Strategy Planner** acts as the mathematical and logical core. It calculates the Best Alternative to a Negotiated Agreement (BATNA). If Supplier A demands \$12 per unit, the Strategy Planner queries the market for Supplier B's spot price. If Supplier B's price is \$14, the agent knows its BATNA is weak and must negotiate cautiously. If Supplier B's price is \$10, the agent can take a more aggressive stance. The Strategy Planner utilizes frontier models like Gemini 3.5 Pro to reason through game-theoretic scenarios, deciding whether to anchor low, concede on volume to win on price, or walk away.

The **Communication Interface** translates the mathematical strategy into natural language. It ingests inbound emails from suppliers, uses structured output extraction to pull out the proposed price, lead time, and terms, and updates the state. It then drafts the outbound counteroffer, ensuring the tone is professional, persuasive, and aligned with corporate communication guidelines.

Finally, the **Evaluation Critic** acts as an internal safeguard. Before any message is transmitted to the external supplier, the Critic reviews the proposed draft against a strict set of policy guardrails. Does this offer exceed the maximum authorized budget? Does it violate compliance standards? If the Critic flags a violation, the draft is rejected and sent back to the Strategy Planner for revision.

1.44.2 Multi-Variable Optimization and Utility Functions

Human negotiations rarely hinge on price alone. A lower unit cost is useless if the components arrive three weeks after the manufacturing line has halted. Agentic procurement systems excel because they can simultaneously optimize across multiple variables—price, lead time, quality SLAs, payment terms, and even Scope 3 carbon emissions—using a defined Utility Function.

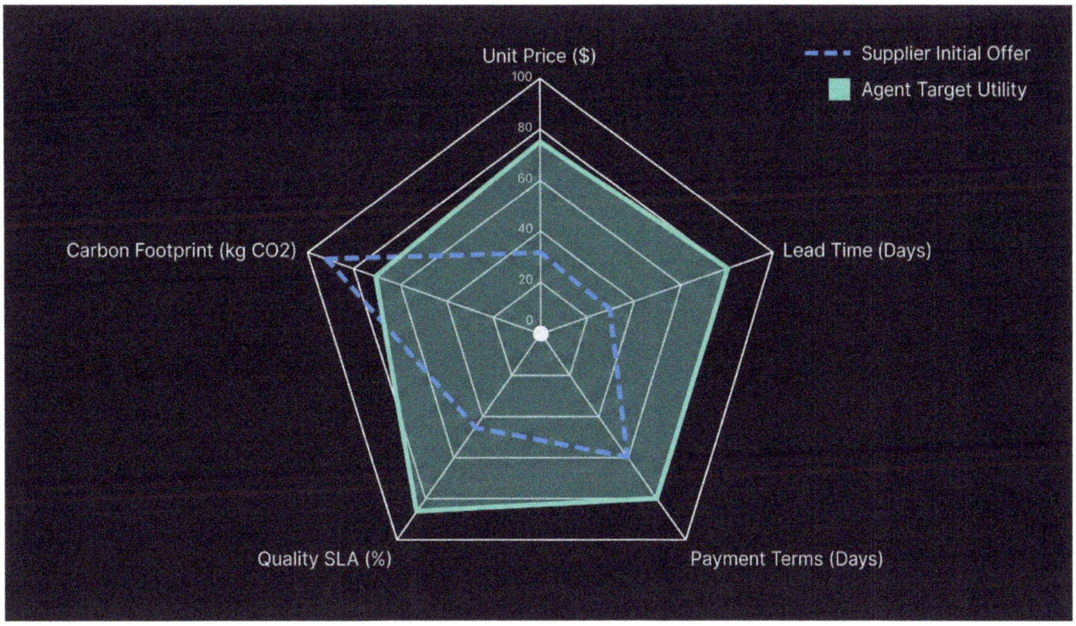

Figure 1.28: The agent evaluates supplier offers not just on price, but across a multi-dimensional utility surface, allowing it to trade concessions (e.g., accepting a higher price for faster lead time) to maximize overall corporate value.

We define the overall utility U of a proposed contract as a weighted sum of its normalized components:

$$U = w_p \cdot N(P) + w_t \cdot N(T) + w_q \cdot N(Q) + w_c \cdot N(C) \tag{1.1}$$

Where P is price, T is lead time, Q is quality, and C is carbon footprint. The weights (w) are dynamically assigned by the S&OP Orchestrator based on the current macroeconomic context. During a supply chain crisis, w_t (time) might be weighted at 0.6, while w_p (price) is reduced to 0.2. During a period of margin compression, the weights invert.

When the agent receives an offer, it calculates the utility score. If $U_{offer} < U_{target}$, the agent must generate a counter-offer. The LLM is instructed to find the path of least resistance to increase U. If historical data indicates that this specific supplier is highly sensitive to payment terms (e.g., they have cash flow issues), the agent might offer Net 15 payment terms (improving the supplier's position) in exchange for a 5% reduction in unit price (improving the agent's U score). This level of dynamic, multi-variable horse-trading is impossible with traditional rules-based procurement software.

1.44.3 Implementing the Negotiation State Machine

To build this system, we utilize LangGraph to manage the stateful, multi-turn nature of the negotiation. The state must track the history of offers, the current round number, the calculated BATNA, and the internal utility scores.

The following code demonstrates the core state graph for the Procurement Negotiation Agent. We define a `NegotiationState` and the nodes responsible for analyzing inbound offers and generating strategic counter-offers.

Listing 1.30: LangGraph Implementation of the Procurement Negotiation Agent

```python
from typing import TypedDict, Annotated, List, Dict, Any
from langgraph.graph import StateGraph, END
from langchain_core.messages import SystemMessage, HumanMessage, AIMessage
from langchain_google_vertexai import ChatVertexAI
from pydantic import BaseModel, Field
import operator

# 1. Define the Structured Output for Offer Extraction
class SupplierOffer(BaseModel):
    unit_price: float = Field(description="The proposed price per unit in USD.")
    lead_time_days: int = Field(description="The proposed delivery time in days.")
    payment_terms: str = Field(description="The proposed payment terms, e.g., Net 30.")
    is_final_offer: bool = Field(description="True if the supplier stated this is their final offer.")

# 2. Define the Negotiation State
class NegotiationState(TypedDict):
    supplier_id: str
    material_id: str
    target_price: float
    max_price: float
    max_lead_time: int
    current_round: int
    max_rounds: int
    message_history: Annotated[List[Any], operator.add]
    latest_offer: SupplierOffer | None
    negotiation_status: str # "IN_PROGRESS", "AGREEMENT_REACHED", "WALK_AWAY", "ESCALATE"

llm = ChatVertexAI(model_name="gemini-3.5-pro", temperature=0.2)
extractor_llm = llm.with_structured_output(SupplierOffer)

# 3. Node: Analyze Inbound Supplier Email
```

```python
def analyze_offer_node(state: NegotiationState):
    # In a real system, the latest message is fetched via an email API integration
    latest_email = state["message_history"][-1].content

    # Extract structured data from the unstructured email
    extracted_offer = extractor_llm.invoke([
        SystemMessage(content="Extract the exact terms from the supplier's email."),
        HumanMessage(content=latest_email)
    ])

    return {"latest_offer": extracted_offer}

# 4. Node: Evaluate Strategy and Utility
def evaluate_strategy_node(state: NegotiationState):
    offer = state["latest_offer"]

    # Hard constraints check
    if offer.unit_price > state["max_price"] and offer.is_final_offer:
        return {"negotiation_status": "WALK_AWAY"}

    if offer.unit_price <= state["target_price"] and offer.lead_time_days <= state["max_lead_time"]:
        return {"negotiation_status": "AGREEMENT_REACHED"}

    if state["current_round"] >= state["max_rounds"]:
        return {"negotiation_status": "ESCALATE"}

    return {"negotiation_status": "IN_PROGRESS"}

# 5. Node: Generate Counter Offer
def generate_counter_node(state: NegotiationState):
    offer = state["latest_offer"]

    system_prompt = f"""You are a senior procurement negotiator for a manufacturing firm.
    Target Price: ${state['target_price']}. Max Price: ${state['max_price']}.
    Supplier currently offered: ${offer.unit_price} with {offer.lead_time_days} days lead time.
    Draft a professional email countering their offer.
    Strategy: Acknowledge their position, but push for a price closer to our target.
    Offer faster payment terms (Net 15 instead of Net 30) if they reduce the price by 5%."""

    messages = [SystemMessage(content=system_prompt)] + state["message_history"]
    response = llm.invoke(messages)

    return {
        "message_history": [AIMessage(content=response.content)],
        "current_round": state["current_round"] + 1
    }
```

```python

# 6. Routing Logic
def route_negotiation(state: NegotiationState):
    status = state["negotiation_status"]
    if status == "IN_PROGRESS":
        return "GenerateCounter"
    elif status in ["AGREEMENT_REACHED", "WALK_AWAY", "ESCALATE"]:
        return END

# 7. Build Graph
workflow = StateGraph(NegotiationState)
workflow.add_node("AnalyzeOffer", analyze_offer_node)
workflow.add_node("EvaluateStrategy", evaluate_strategy_node)
workflow.add_node("GenerateCounter", generate_counter_node)

workflow.set_entry_point("AnalyzeOffer")
workflow.add_edge("AnalyzeOffer", "EvaluateStrategy")
workflow.add_conditional_edges("EvaluateStrategy", route_negotiation)
workflow.add_edge("GenerateCounter", END) # In reality, this waits for the
    next supplier email

negotiation_app = workflow.compile()
```

This architecture ensures that the LLM is not just blindly generating text. It is constrained by a deterministic state machine. The `extractor_llm` forces the unstructured email into a strict Pydantic schema (`SupplierOffer`). The `evaluate_strategy_node` uses standard Python logic—not probabilistic generation—to check if the offer violates the hard `max_price` ceiling. Only if the deterministic checks pass does the system invoke the LLM to generate the natural language counter-offer.

1.44.4 Sequence and Protocol Integration

In a production environment, the agent does not operate in a vacuum. It must interface with standard enterprise communication protocols. The most common deployment pattern involves connecting the LangGraph application to a dedicated procurement email inbox (e.g., `sourcing-agent@company.com`) via Microsoft Graph API or Google Workspace APIs.

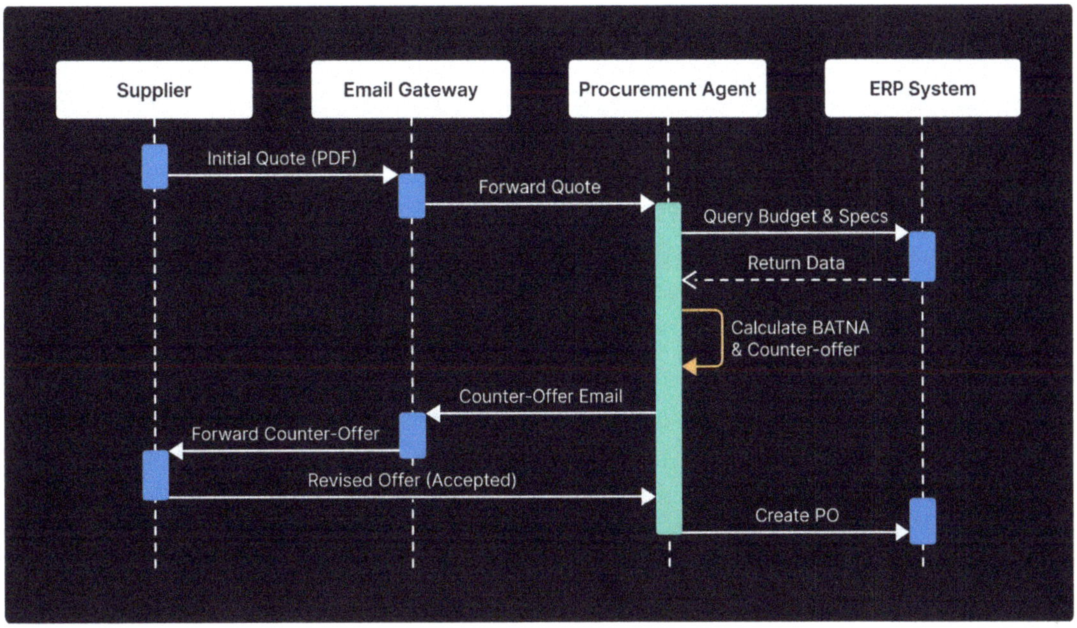

Figure 1.29: Sequence diagram detailing the asynchronous, event-driven communication flow between the external supplier, the agent, and internal enterprise systems.

When an email arrives, a webhook triggers the LangGraph execution. The agent processes the message, updates its state, and drafts a reply. However, to prevent runaway loops or inappropriate communications, enterprise deployments heavily utilize the Human-in-the-Loop (HITL) pattern.

Before the `GenerateCounter` node actually dispatches the email via the API, the draft is routed to a human buyer's dashboard. The dashboard displays the supplier's original email, the agent's extracted data (price, terms), the agent's calculated utility score, and the proposed draft response. The human buyer can click "Approve and Send," "Edit," or "Take Over." As the agent proves its reliability over hundreds of supervised negotiations, the organization can lower the HITL threshold, allowing the agent to autonomously negotiate contracts under a specific dollar value (e.g., <$50,000), while routing larger contracts for human review.

Once an agreement is reached (`AGREEMENT_REACHED` state), the agent utilizes the Agent Payments Protocol (AP2) or standard ERP APIs to finalize the transaction. It automatically generates the Purchase Order (PO) in SAP, attaches the negotiated email thread as an audit trail, and routes the PO to the supplier for signature.

1.44.5 Performance and Capability Comparison

The transition from traditional procurement automation to agentic negotiation represents a step-function increase in capability. Traditional systems are limited to structured data and predefined rules. Agentic systems can handle ambiguity, unstructured text, and dynamic strategy formulation.

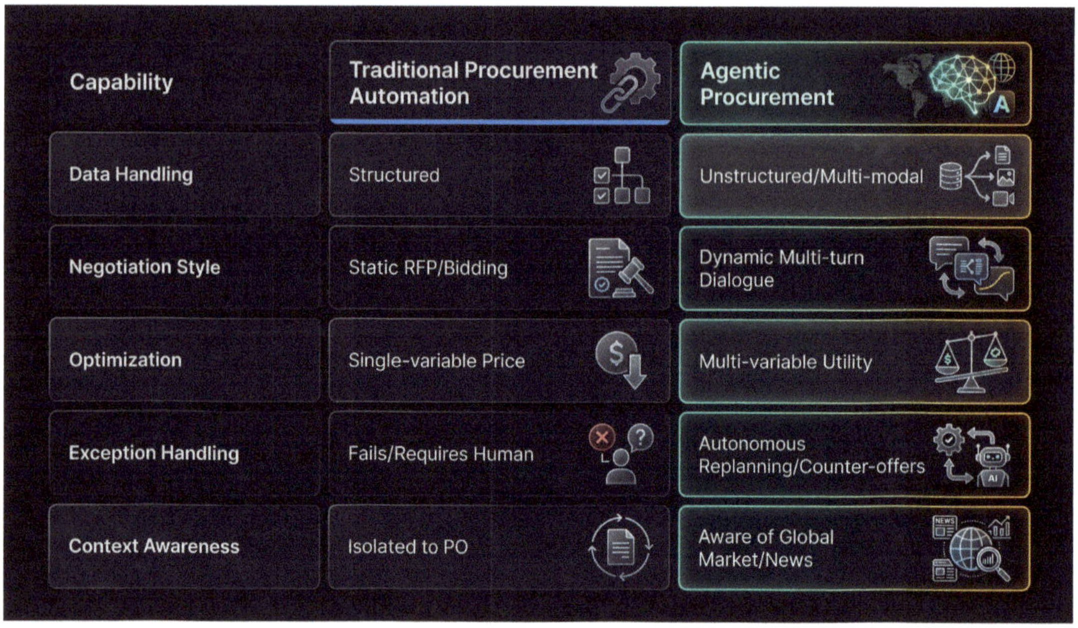

Figure 1.30: Capability matrix highlighting the paradigm shift from rules-based automation to autonomous agentic negotiation in enterprise procurement.

In benchmark deployments within the manufacturing sector, Procurement Negotiation Agents have demonstrated a 15-20% reduction in cycle time for direct materials sourcing. Because the agent can conduct negotiations concurrently with hundreds of suppliers 24/7, it eliminates the bottleneck of human buyer bandwidth. Furthermore, by strictly adhering to the multi-variable utility function, agents consistently secure contracts that offer better overall corporate value, avoiding the human cognitive bias of over-focusing on unit price at the expense of lead time or quality.

However, these systems are highly sensitive to token limits and context window degradation. A negotiation that spans 40 emails over three weeks can easily exceed standard context windows or become diluted, causing the LLM to "forget" a concession made in week one. To mitigate this, the `NegotiationState` must implement context compression. Instead of passing the raw `message_history` to the LLM every time, a background summarization agent periodically compresses older emails into a dense "Negotiation Summary" string, preserving the core facts (agreed terms, remaining sticking points) while freeing up token space for the immediate tactical reasoning required for the next counter-offer.

Key Takeaways

- **Multi-Variable Utility:** Agentic procurement transcends simple price comparison, optimizing across price, lead time, terms, and ESG metrics using a dynamic utility function.
- **Stateful Orchestration:** Negotiations are multi-turn processes. Frameworks like LangGraph are essential for maintaining state, tracking rounds, and enforcing deterministic constraints (like max budget) outside the LLM.

- **Structured Extraction is Critical:** The agent must convert unstructured supplier emails into structured data (Pydantic schemas) to perform mathematical BATNA calculations and policy checks.
- **Context Compression:** Long-running negotiations require active memory management. Summarize historical turns to prevent context window overflow and maintain strategic focus.
- **Progressive Autonomy:** Deploy with strict Human-in-the-Loop (HITL) dashboards. Gradually increase the autonomous execution threshold (e.g., PO value limits) as the agent's reliability is proven in production.

Learning Objectives

- Understand the limitations of traditional Material Requirements Planning (MRP) and static min-max inventory models in volatile supply chain environments.
- Architect an agentic replenishment system that fuses structured ERP data with unstructured supplier communications and external market signals.
- Implement a stateful inventory planner using LangGraph, integrating deterministic mathematical models (like EOQ) with probabilistic LLM reasoning.
- Design multi-objective utility functions that allow agents to autonomously balance holding costs, stockout risks, and supplier reliability.
- Deploy Human-in-the-Loop (HITL) guardrails for high-value autonomous procurement decisions.

1.45 The Evolution from Static MRP to Agentic Replenishment

For decades, the backbone of manufacturing supply chains has been Material Requirements Planning (MRP) systems and their modern Enterprise Resource Planning (ERP) descendants. These systems rely heavily on deterministic, rules-based logic. The most common paradigm is the Reorder Point (ROP) and Min-Max system, where an order is triggered when inventory falls below a specific threshold, calculated using historical average demand and a static, assumed lead time.

While these systems are highly efficient in stable environments, they are fundamentally brittle in the face of modern supply chain volatility. Traditional MRP operates under the assumption that the future will look exactly like the past. When a supplier quotes a 14-day lead time, the ERP hardcodes that integer. If a geopolitical event, a port strike, or a sudden raw material shortage occurs, the ERP remains entirely oblivious until the shipment fails to arrive, resulting in a stockout, a halted production line, and massive expediting costs.

The core deficiency of traditional systems is their inability to process *context*. Supply chain disruptions rarely manifest first as structured data in an SQL database; they manifest as unstructured text—an apologetic email from a supplier account manager, a news alert about a factory fire in Shenzhen, or a weather bulletin regarding a typhoon in the South China Sea. Traditional systems require human planners to read this unstructured data, synthesize its impact, manually override the ERP parameters, and execute a new plan. In a global manufacturing operation with tens of thousands of SKUs, this manual intervention does not scale, leading to the well-documented "bullwhip effect" where small fluctuations cause massive downstream inventory distortions.

Agentic AI introduces a paradigm shift: the **Inventory Replenishment Planner**. By leveraging Large Language Models (LLMs) orchestrated within a stateful framework like LangGraph or the Google Agent Development Kit (ADK), we can build systems that possess both the mathematical rigor of an ERP and the contextual awareness of a human planner.

An agentic replenishment system does not replace the ERP; it acts as an intelligent orchestration layer above it. It continuously monitors inventory levels via API, but it also ingests supplier emails, reads market news, and queries predictive machine learning models for demand forecasting. When an anomaly is detected—such as a supplier emailing to say, "Due to customs delays, your shipment of microcontrollers will be four days late"—the agent parses the natural language, updates the internal state, explodes the Bill of Materials (BOM) to understand the impact on finished goods, and autonomously formulates a mitigation strategy. This strategy might involve placing an expedited order with a secondary, more expensive local supplier to bridge the gap, thereby preventing a line-down situation.

This transition from reactive, rules-based automation to proactive, goal-seeking autonomy represents one of the highest ROI applications of Agentic AI in the Industry 4.0 landscape. It directly reduces safety stock requirements (freeing up working capital) while simultaneously increasing service levels and supply chain resilience.

1.46 Architecture of the Replenishment Agent

Designing an Inventory Replenishment Planner requires a careful architectural balance. We must strictly separate deterministic operations (like calculating the Economic Order Quantity or querying current stock levels) from probabilistic operations (like interpreting a supplier's tone in an email or reasoning about geopolitical risks). Forcing an LLM to do arithmetic is an anti-pattern; instead, the LLM should act as the reasoning engine that decides *which* mathematical tools to invoke and *how* to interpret their outputs.

The architecture of a production-grade replenishment agent consists of four primary layers:

1. The Integration and Perception Layer: This layer acts as the agent's sensory organs. It includes standard API connectors to the ERP system (e.g., SAP S/4HANA via OData or BAPI) to read real-time inventory levels, open purchase orders, and BOM structures. Crucially, it also includes unstructured data connectors: Microsoft Graph or Google Workspace APIs to monitor dedicated procurement email inboxes, and web search tools (like Google Search API) to monitor macro-environmental factors.

2. The Tool Registry (Actuators and Calculators): The agent is equipped with a suite of deterministic tools. These include:

- `calculate_eoq(annual_demand, order_cost, holding_cost)`: A standard Python function that returns the mathematically optimal order quantity.
- `get_demand_forecast(sku_id, horizon_days)`: A tool that queries a dedicated time-series forecasting model (e.g., deployed on Vertex AI) rather than relying on the LLM to guess future demand.
- `check_supplier_contracts(supplier_id)`: A Retrieval-Augmented Generation (RAG) tool that searches the enterprise knowledge base to retrieve SLA terms, penalty

clauses, and alternative sourcing agreements.

3. The Cognitive Orchestration Layer (LangGraph): This is the brain of the system. We utilize a state graph to enforce a rigorous, repeatable workflow. The state object maintains the context of the current replenishment cycle (SKU, current stock, forecasted demand, identified risks, proposed actions). The graph routes the execution through specific nodes: data gathering, risk assessment, strategy formulation, and execution.

4. The Governance and Execution Layer: Before any action is taken in the real world (e.g., committing a $100,000 Purchase Order), the system must pass through governance gates. For low-value, low-risk items (e.g., standard fasteners), the agent may have full autonomy to execute the PO via the ERP API. For critical components or high-value orders, the agent generates a comprehensive "Decision Memo" and routes it to a human planner's dashboard for a single-click approval (Human-in-the-Loop).

This architecture ensures that the LLM is grounded in enterprise reality. It cannot hallucinate inventory levels because it must retrieve them via the ERP tool. It cannot invent a supplier because it must select from the approved vendor list. The LLM's sole responsibility is to apply human-like reasoning to the synthesis of these disparate data streams.

1.47 Implementation: Building the Planner with LangGraph

To demonstrate the mechanics of this pattern, we will implement the core logic of the Inventory Replenishment Planner using LangGraph and the Google Vertex AI SDK (Gemini 3.5 Pro). The implementation relies on a strictly typed state object to ensure data consistency across the reasoning cycle.

First, we define the state and the structured output schemas. We use Pydantic to force the LLM to output its final replenishment plan in a machine-readable format, which is critical for downstream API execution.

Listing 1.31: Defining the State and Schemas for the Replenishment Agent

```
from typing import TypedDict, Annotated, List, Dict, Any, Optional
from pydantic import BaseModel, Field
import operator

# 1. Define the Structured Output for the Final Plan
class ReplenishmentAction(BaseModel):
    sku_id: str = Field(description="The ID of the SKU to replenish.")
    primary_supplier_id: str = Field(description="The ID of the chosen supplier.")
    order_quantity: int = Field(description="The exact quantity to order.")
    expedite_shipping: bool = Field(description="Whether to pay for expedited shipping.")
    rationale: str = Field(description="Detailed reasoning for this decision.")
    requires_human_approval: bool = Field(description="True if order value > $50k or high risk.")

# 2. Define the Graph State
class ReplenishmentState(TypedDict):
```

```
    sku_id: str
    current_inventory: int
    safety_stock_level: int
    forecasted_demand_30d: int
    supplier_communications: List[str] # Recent emails from suppliers
    identified_risks: List[str]
    optimal_eoq: int
    proposed_action: Optional[ReplenishmentAction]
    execution_status: str
```

Next, we define the nodes of our state graph. Each node represents a distinct phase in the agent's lifecycle: Sense, Plan, Act, and Reflect. In this simplified example, we mock the external API calls (ERP, Forecasting) that would normally be executed via LangChain tools.

Listing 1.32: LangGraph Nodes for Data Gathering and Reasoning

```
from langchain_core.messages import SystemMessage, HumanMessage
from langchain_google_vertexai import ChatVertexAI
from langgraph.graph import StateGraph, END

llm = ChatVertexAI(model_name="gemini-3.5-pro", temperature=0.1)
structured_llm = llm.with_structured_output(ReplenishmentAction)

# Node 1: Gather Deterministic Data (ERP & ML Forecasts)
def gather_data_node(state: ReplenishmentState):
    # In production, these would be API calls to SAP and Vertex AI
    sku = state["sku_id"]

    # Mock data retrieval
    current_inv = 1200
    safety_stock = 500
    forecast = 2500 # High demand expected
    eoq = 1000 # Calculated via standard formula

    # Mock unstructured data (e.g., fetched from Microsoft Graph API)
    emails = [
        "From Supplier A: Please note that due to port congestion, all
    shipments next week will face a 5-day delay."
    ]

    return {
        "current_inventory": current_inv,
        "safety_stock_level": safety_stock,
        "forecasted_demand_30d": forecast,
        "optimal_eoq": eoq,
        "supplier_communications": emails
    }

# Node 2: Assess Risks (LLM Reasoning over Unstructured Data)
def assess_risks_node(state: ReplenishmentState):
    prompt = f"""
    Analyze the following supply chain data for SKU {state['sku_id']}.
```

```
    Current Inventory: {state['current_inventory']}
    Forecasted 30-day Demand: {state['forecasted_demand_30d']}
    Supplier Emails: {state['supplier_communications']}

    Identify any risks to fulfilling the forecasted demand.
    Return a concise list of risk factors.
    """
    response = llm.invoke([HumanMessage(content=prompt)])
    # Simple parsing for demonstration
    risks = response.content.split('\n')
    return {"identified_risks": risks}

# Node 3: Formulate Replenishment Strategy
def formulate_strategy_node(state: ReplenishmentState):
    prompt = f"""
    You are an expert Inventory Replenishment Agent.
    SKU: {state['sku_id']}
    Inventory: {state['current_inventory']} (Safety Stock: {state['
    safety_stock_level']})
    30-Day Demand: {state['forecasted_demand_30d']}
    Standard EOQ: {state['optimal_eoq']}
    Identified Risks: {state['identified_risks']}

    Formulate a replenishment plan. If there are shipping delays, you may need
     to
    order more than the EOQ or expedite shipping to prevent a stockout.
    If the total order value (Quantity * $50/unit) exceeds $50,000, flag for
    human approval.
    """

    action = structured_llm.invoke([
        SystemMessage(content="You are a supply chain optimization engine."),
        HumanMessage(content=prompt)
    ])

    return {"proposed_action": action}
```

The critical design pattern here is the separation of concerns. The `gather_data_node` is entirely deterministic; it fetches hard numbers. The `assess_risks_node` uses the LLM's natural language understanding to parse the supplier email and realize that a "5-day delay" combined with high forecasted demand creates a stockout risk. Finally, the `formulate_strategy_node` synthesizes the math and the risk assessment to generate a structured `ReplenishmentAction`.

To complete the system, we define the routing logic and compile the graph. The routing logic enforces our governance model: if the LLM flagged the order as requiring human approval, the graph routes to a HITL node; otherwise, it routes directly to execution.

Listing 1.33: Routing Logic and Graph Compilation

```
# Node 4: Human-in-the-Loop (HITL)
def human_review_node(state: ReplenishmentState):
    # In production, this pauses execution and sends an alert to a UI
```

```
    dashboard
    print(f"ALERT: Human review required for SKU {state['sku_id']}.")
    print(f"Rationale: {state['proposed_action'].rationale}")
    # Mocking human approval
    return {"execution_status": "APPROVED_BY_HUMAN"}

# Node 5: Execute Purchase Order
def execute_po_node(state: ReplenishmentState):
    action = state["proposed_action"]
    # In production, this makes an API call to SAP to create the PO
    print(f"Executing PO for {action.order_quantity} units from {action.
    primary_supplier_id}.")
    return {"execution_status": "PO_CREATED"}

# Routing Function
def route_execution(state: ReplenishmentState):
    if state["proposed_action"].requires_human_approval:
        return "HumanReview"
    return "ExecutePO"

# Build and Compile the Graph
workflow = StateGraph(ReplenishmentState)

workflow.add_node("GatherData", gather_data_node)
workflow.add_node("AssessRisks", assess_risks_node)
workflow.add_node("FormulateStrategy", formulate_strategy_node)
workflow.add_node("HumanReview", human_review_node)
workflow.add_node("ExecutePO", execute_po_node)

workflow.set_entry_point("GatherData")
workflow.add_edge("GatherData", "AssessRisks")
workflow.add_edge("AssessRisks", "FormulateStrategy")
workflow.add_conditional_edges("FormulateStrategy", route_execution)
workflow.add_edge("HumanReview", "ExecutePO")
workflow.add_edge("ExecutePO", END)

replenishment_app = workflow.compile()
```

1.48 Dynamic Disruption Handling and Multi-Objective Optimization

The true value of the Inventory Replenishment Planner emerges during supply chain disruptions. Consider a scenario where a manufacturer relies on a primary supplier in Taiwan for a critical component. A sudden geopolitical event or natural disaster occurs.

A traditional MRP system will continue to generate purchase orders to the Taiwanese supplier based on the static lead time of 14 days, completely unaware that the port is closed. The inventory will slowly drain until a stockout occurs, halting the entire production line.

An agentic system handles this entirely differently. Through its integration with news APIs and supplier email monitoring, the agent detects the disruption in real-time. The `assess_risks_-`

node flags the primary supplier as compromised and dynamically adjusts the expected lead time from 14 days to "unknown/infinite."

At this point, the agent engages in **Multi-Objective Optimization**. It must balance three competing variables: 1. **Stockout Risk (Service Level):** The cost of halting the production line (often tens of thousands of dollars per hour). 2. **Procurement Cost:** The price of the component. 3. **Holding Cost:** The cost of storing excess inventory.

The agent queries its tool registry for alternative suppliers. It finds a secondary supplier in Mexico (nearshoring). The Mexican supplier charges a 30% premium per unit but has a guaranteed lead time of 3 days. The agent calculates the utility function: the 30% premium is vastly cheaper than the cost of a line-down event.

The agent formulates a new strategy: it splits the order. It places an immediate, expedited order with the Mexican supplier for just enough inventory to cover the next 14 days of forecasted demand (minimizing the premium paid), while placing a standard order with the Taiwanese supplier for the remainder of the EOQ, assuming the port will reopen eventually. It then generates a detailed rationale and routes this complex, high-value decision to the Human-in-the-Loop dashboard.

1.48.1 Performance, Latency, and Cost Considerations

Deploying this pattern at enterprise scale requires careful attention to token economics and API rate limits. A large manufacturer may have 50,000 active SKUs. Running a full LLM reasoning cycle for every SKU every night is computationally prohibitive and largely unnecessary, as 90% of SKUs will experience normal, stable demand.

The best practice is to implement a **Triage Architecture**. A lightweight, deterministic script (or a small, fast model like Gemini 3.5 Flash-Lite) runs nightly across all 50,000 SKUs. It looks for simple triggers: inventory dropping below safety stock, sudden spikes in the ML demand forecast, or new unread emails from the supplier associated with that SKU.

Only if a trigger is hit does the system invoke the heavy, reasoning-intensive LangGraph workflow powered by a frontier model like Gemini 3.5 Pro. This reduces token consumption by orders of magnitude while ensuring that complex cognitive power is applied exactly where it is needed: managing exceptions and disruptions.

In benchmark deployments, organizations utilizing this triage-to-agent architecture have reported a 15-20% reduction in required safety stock (as the system reacts faster to disruptions, reducing the need for massive buffers) and a near-elimination of stockouts caused by missed supplier communications. The latency for a single SKU reasoning cycle typically ranges from 5 to 12 seconds, making it highly suitable for asynchronous, batch-oriented nightly MRP runs.

Key Takeaways

- **Context is King:** Traditional MRP fails because it cannot process unstructured context. Agentic planners bridge this gap by reading emails, news, and contracts to dynamically adjust lead times and risk profiles.
- **Separate Math from Reasoning:** Never ask an LLM to calculate an EOQ or a safety stock level. Use deterministic tools (Python functions, ERP APIs) for math, and use the LLM to reason about the outputs of those tools.
- **Stateful Orchestration:** Use frameworks like LangGraph to enforce a strict Sense → Plan → Act loop, ensuring the agent gathers all necessary data before formulating a strategy.
- **Triage for Scale:** Do not run heavy LLM reasoning on every SKU every day. Use deterministic triggers to identify exceptions, and route only those exceptions to the agentic planner to optimize token costs and latency.
- **Multi-Objective Utility:** Empower the agent to make trade-offs (e.g., paying a premium for nearshoring to avoid a line-down event) by defining clear utility functions and business rules in the system prompt.

Learning Objectives

- Understand the limitations of static routing guides in traditional Transportation Management Systems (TMS) and how agentic AI overcomes them.
- Architect a multi-agent system capable of dynamically selecting freight modes (air, ocean, rail, truck) based on real-time constraints, unstructured disruption data, and multi-objective utility functions.
- Implement a context-aware routing workflow using LangGraph that balances procurement cost, transit time, risk, and Scope 3 carbon emissions.
- Design integration patterns between the agentic mode selector, enterprise ERP systems (like SAP), and external market data APIs (spot rates, weather, port congestion).
- Evaluate the token economics, latency trade-offs, and human-in-the-loop governance required for deploying autonomous freight booking at enterprise scale.

1.49 The Paradigm Shift in Freight Routing

In the previous chapter, we explored how the Inventory Replenishment Planner dynamically calculates optimal order quantities and safety stock levels in response to supply chain disruptions. However, generating a purchase order is only half the battle. Once the inventory is ready at the supplier's dock, the organization must decide *how* to move that freight to the manufacturing facility or distribution center. This decision—the selection of the transportation mode and the specific carrier—is traditionally governed by a Transportation Management System (TMS) relying on static routing guides. In the era of Industry 4.0 and highly volatile global supply chains, this static approach is no longer sufficient, necessitating a paradigm shift toward agentic, context-aware transportation mode selection.

Traditional TMS platforms operate on rigid, rule-based logic. A typical routing guide might dictate: "If the shipment is under 150 lbs, use Parcel Carrier A; if it is between 150 and 10,000 lbs, use LTL (Less-Than-Truckload) Carrier B; if it is international from Asia to North America, use Ocean Freight Carrier C unless the priority flag is set to 'High', in which case use Air Freight Carrier D." These rules are usually negotiated annually during procurement bids and hardcoded into the ERP or TMS. While this deterministic approach is computationally cheap and easy to audit, it is fundamentally blind to context. It assumes a static world where transit times are guaranteed, spot market rates do not fluctuate, ports do not experience labor strikes, and weather patterns do not disrupt rail networks.

The reality of modern manufacturing logistics is entirely different. Consider a Just-In-Time (JIT) automotive assembly line. If a shipment of microchips from Taiwan is delayed by a typhoon, the static routing guide will still default to ocean freight because the shipment weight and standard priority code dictate it. The TMS does not know that the assembly line will shut down in exactly 14 days if those chips do not arrive, incurring a "line-down" cost of $50,000 per hour. A human logistics planner must manually intervene, override the TMS, and book expedited air freight. Conversely, if a shipment is marked "High Priority" but the production schedule has shifted and the parts are not actually needed for 60 days, the static rule will blindly book expensive air freight, wasting tens of thousands of dollars when standard ocean freight would have sufficed.

Agentic AI introduces dynamic, context-aware decision-making to freight routing. An Agentic Transportation Mode Selector does not merely execute a rule; it evaluates the holistic state of the supply chain. It ingests the demand signal from the ERP, reads the production schedule, queries real-time spot market APIs for current freight rates, checks global weather and port congestion data, and calculates the exact utility of every possible transportation mode. More importantly, it can process unstructured data—such as an email from a freight forwarder warning of an impending dockworker strike in Long Beach, or a news alert about a blockage in the Suez Canal—and immediately factor that risk into its routing logic.

Furthermore, the modern supply chain is increasingly driven by sustainability mandates. Organizations are under immense pressure to reduce their Scope 3 carbon emissions (emissions from assets not owned or controlled by the reporting organization, which includes third-party logistics). Traditional systems treat carbon tracking as a post-shipment reporting exercise. An agentic system treats carbon as a primary variable in its pre-shipment optimization algorithm. If a shipment is not urgently needed, the agent can autonomously choose a slower, lower-carbon mode (e.g., rail instead of truck, or ocean instead of air) to optimize the company's carbon budget, dynamically trading off transit time for sustainability.

The transition to an agentic mode selector requires a fundamental shift in how we architect logistics software. We move from a paradigm of "If X, then Y" to a paradigm of "Given state S, constraints C, and objectives O, formulate and execute the optimal plan P." This requires the integration of Large Language Models (LLMs) for reasoning over unstructured data, deterministic APIs for fetching real-time market conditions, and robust state machines to orchestrate the decision-making process safely and reliably. In the following sections, we will dissect the architecture required to build this capability, utilizing frameworks like LangGraph and the Google Gemini Enterprise stack to ensure enterprise-grade scalability and governance.

1.50 Architecture of the Agentic Mode Selector

Building a robust Agentic Transportation Mode Selector requires a sophisticated architecture that blends deterministic data retrieval with non-deterministic reasoning. We cannot rely on an LLM to simply "guess" the best shipping route based on a text prompt; the financial and operational stakes are too high. Instead, we must architect a multi-agent workflow where the LLM acts as an orchestrator and reasoning engine, interacting with a suite of specialized tools and sub-agents to gather facts, calculate costs, and assess risks before making a final recommendation.

The architecture is built upon a stateful graph, typically implemented using LangGraph or the Google Agent Development Kit (ADK). The state object (e.g., `ShipmentState`) acts as the central memory for the transaction, accumulating data as the shipment moves through various nodes of evaluation. The core components of this architecture include the Data Ingestion Layer, the Context Engineering Engine, the Multi-Agent Evaluation Swarm, and the Execution Gateway.

1. Data Ingestion and Context Engineering Layer: The process begins when a shipment requirement is generated by the ERP (e.g., SAP or Oracle) or the Inventory Replenishment Planner discussed in the previous chapter. This requirement includes basic metadata: Origin, Destination, Weight, Volume, Ready Date, and Required Delivery Date (RDD). However, basic metadata is insufficient for agentic reasoning. The Context Engineering Engine enriches this request by querying adjacent enterprise systems. It queries the Manufacturing Execution System (MES) to determine the exact production order this shipment supports. It queries the financial system to determine the value of the goods and the cost of a potential line-down event. This enriched context transforms a generic request ("Ship 500kg from Shanghai to Detroit") into a highly specific operational imperative ("Ship 500kg of critical microchips from Shanghai to Detroit; if they do not arrive by October 15th, Assembly Line 3 will halt, costing $120,000 per day").

2. The Multi-Agent Evaluation Swarm: Once the context is established, the Orchestrator Agent delegates the evaluation to a swarm of specialized sub-agents or tool-calling nodes.

- **The Market Rate Agent:** This node interfaces with external freight APIs (e.g., Freightos, DAT, or direct carrier APIs) to pull real-time spot rates and transit times for various modes (Air, Ocean, Rail, Truck). It also retrieves contracted rates from the internal TMS database.
- **The Risk & Disruption Agent:** This is where the LLM's natural language capabilities shine. This agent monitors unstructured data streams—global news feeds, weather alerts, AIS vessel tracking anomalies, and supplier emails. It uses a frontier model like Gemini 3.5 Pro to read these streams and identify risks specific to the proposed lanes. For example, it might read a news report about a strike at the Port of Vancouver and flag the Ocean/Rail multimodal option as "High Risk," dynamically increasing its projected transit time.
- **The Sustainability Agent:** This node calculates the estimated Scope 3 CO2 equivalent emissions for each proposed route and mode, utilizing standard frameworks like the GLEC (Global Logistics Emissions Council) framework.

3. The Execution Gateway and Human-in-the-Loop (HITL): After the swarm gathers the data, the Orchestrator synthesizes the findings and calculates a multi-objective utility score for each option (detailed in the next section). The system then selects the optimal mode and carrier. However, autonomous execution in enterprise logistics requires strict governance. The Execution Gateway evaluates the agent's decision against predefined business rules. If the selected mode is standard, within budget, and low risk, the agent makes an API call to the TMS to automatically book the freight. If the agent selects an expedited mode that exceeds a certain cost threshold (e.g., booking a $40,000 air charter), or if the confidence score of the risk assessment is low, the workflow pauses. It packages the context, the options evaluated, and the agent's rationale into a concise summary and routes it to a human logistics planner for approval.

The following code listing demonstrates how this orchestration can be implemented using LangGraph. We define a state object to hold the shipment details and the gathered quotes, and we create nodes for data gathering, risk assessment, and final decision-making.

Listing 1.34: LangGraph Implementation of the Mode Selector Workflow

```python
from typing import TypedDict, List, Dict, Any
from langgraph.graph import StateGraph, END
from langchain_google_vertexai import ChatVertexAI
from langchain_core.messages import HumanMessage, SystemMessage
from pydantic import BaseModel, Field

# 1. Define the State and Structured Output
class RouteOption(BaseModel):
    mode: str
    carrier: str
    cost_usd: float
    transit_days: int
    co2_emissions_kg: float
    risk_score: int = Field(description="Risk score from 1-100")

class ShipmentState(TypedDict):
    shipment_id: str
    origin: str
    destination: str
    weight_kg: float
    required_delivery_date: str
    line_down_cost_per_day: float
    available_routes: List[RouteOption]
    unstructured_intel: str
    selected_route: RouteOption
    rationale: str
    requires_human_approval: bool

# Initialize the LLM (Gemini 3.5 Pro for complex reasoning)
llm = ChatVertexAI(model_name="gemini-3.5-pro", temperature=0.1)

# 2. Node: Gather Market Data (Mocked API Call)
def gather_market_data(state: ShipmentState):
    # In production, this calls Freightos, DAT, or internal TMS APIs
    routes = [
        RouteOption(mode="Ocean", carrier="Maersk", cost_usd=1200,
```

```
transit_days=35, co2_emissions_kg=450, risk_score=10),
        RouteOption(mode="Air", carrier="FedEx", cost_usd=8500, transit_days=4, co2_emissions_kg=4200, risk_score=5),
        RouteOption(mode="Sea-Air", carrier="Expeditors", cost_usd=4100, transit_days=18, co2_emissions_kg=1800, risk_score=15)
    ]
    return {"available_routes": routes}

# 3. Node: Assess Risk via Unstructured Data
def assess_risk(state: ShipmentState):
    # In production, this fetches recent news/emails regarding the origin/destination
    intel = "News Alert: Port workers in Los Angeles have announced a 3-day strike starting next week. Heavy congestion expected."

    prompt = f"""
    Analyze the following intelligence: '{intel}'
    Adjust the risk scores (1-100) for the following routes based on this intel.
    Routes: {[r.dict() for r in state['available_routes']]}
    Return ONLY the updated list of routes in JSON format.
    """
    # Simplified for demonstration; in reality, use structured output parsing
    # Assume the LLM increases the Ocean risk score to 85 due to the LA port strike
    updated_routes = state['available_routes']
    updated_routes.risk_score = 85

    return {"unstructured_intel": intel, "available_routes": updated_routes}

# 4. Node: Multi-Objective Decision Engine
def select_optimal_mode(state: ShipmentState):
    prompt = f"""
    You are an expert Logistics Orchestrator. Select the optimal shipping route.
    Shipment ID: {state['shipment_id']}
    Required Delivery Date: {state['required_delivery_date']}
    Line-Down Cost: ${state['line_down_cost_per_day']}/day

    Available Routes:
    {[r.dict() for r in state['available_routes']]}

    Intelligence: {state['unstructured_intel']}

    Calculate the total utility considering freight cost, line-down risk, and carbon emissions.
    Provide the selected route, a detailed rationale, and a boolean flag indicating if human approval is needed (True if cost > $5000 or risk > 50).
    """

    # Use structured output to guarantee schema adherence
    structured_llm = llm.with_structured_output(
        schema={"type": "object", "properties": {"selected_mode": {"type": "string"}, "rationale": {"type": "string"}, "requires_human_approval": {"
```

```
    type": "boolean"}}}
    )

    decision = structured_llm.invoke([SystemMessage(content="You are a logistics AI."), HumanMessage(content=prompt)])

    # Match selected mode back to the object
    selected = next(r for r in state['available_routes'] if r.mode == decision["selected_mode"])

    return {
        "selected_route": selected,
        "rationale": decision["rationale"],
        "requires_human_approval": decision["requires_human_approval"]
    }

# 5. Compile the Graph
workflow = StateGraph(ShipmentState)
workflow.add_node("GatherData", gather_market_data)
workflow.add_node("AssessRisk", assess_risk)
workflow.add_node("SelectMode", select_optimal_mode)

workflow.set_entry_point("GatherData")
workflow.add_edge("GatherData", "AssessRisk")
workflow.add_edge("AssessRisk", "SelectMode")
workflow.add_edge("SelectMode", END)

mode_selector_app = workflow.compile()
```

This architecture ensures that the LLM is not hallucinating freight rates or transit times. It is strictly reasoning over deterministic data provided by the API nodes, applying its semantic understanding only to the unstructured risk intelligence and the final multi-objective trade-off analysis.

1.51 Multi-Objective Optimization and Real-World Constraints

The core cognitive task of the Agentic Mode Selector is performing Multi-Objective Optimization (MOO). In logistics, there is rarely a single "best" answer; there is only the optimal trade-off between competing variables. The primary variables are Procurement Cost (the price paid to the carrier), Transit Time (speed to destination), Reliability/Risk (variance in transit time), and Sustainability (carbon footprint).

Traditional systems attempt to solve this using linear programming or simple weighted scoring models. However, these models fail when the relationships between variables are non-linear or highly contextual. For instance, the value of saving one day of transit time is negligible if the parts are not needed for a month. But if the factory is facing a line-down event tomorrow, the value of saving one day of transit time becomes astronomical.

An agentic system handles this by dynamically generating a utility function based on the specific context of the shipment. The Orchestrator Agent evaluates the utility U of a given route i using a dynamically weighted function:

$$U_i = -(w_c \cdot Cost_i) - (w_t \cdot Penalty(Time_i)) - (w_r \cdot Risk_i) - (w_e \cdot Emissions_i)$$

The brilliance of the agentic approach lies in how the weights (w_c, w_t, w_r, w_e) and the $Penalty$ function are determined. Instead of hardcoding them, the LLM derives them from the context engineered in the first step.

Table~1.3 contrasts how the agent reweights its objectives between a routine replenishment and an aircraft-on-ground (AOG) emergency.

Table 1.3: Dynamic utility weighting by shipment context

Objective	**Standard Replenishment**	**AOG / Critical**
Cost weight (w_c)	High	Low (cost ignored)
Time weight (w_t)	Very low	Maximum
Emissions weight (w_e)	High	Low (emissions ignored)
Delay penalty	Zero (30 vs. 15 days equal)	Exponential per hour of delay
Mode selected	Cheapest, greenest ocean freight	Fastest air charter

In the AOG case—a critical aerospace term meaning a plane cannot fly until the part arrives—the agent ignores cost and emissions entirely, justifying the \$50,000 air-charter premium by citing the \$150,000/day cost of the grounded aircraft.

1.51.1 Handling Unstructured Disruptions: The Red Sea Scenario

To understand the power of this pattern, consider a real-world disruption scenario: the closure of the Red Sea / Suez Canal corridor due to geopolitical conflict.

In a traditional setup, the TMS has a contracted ocean lane from Mumbai to Rotterdam with a stated transit time of 24 days. When the conflict begins, vessels are forced to reroute around the Cape of Good Hope, adding 12-15 days to the journey. The static TMS does not know this. It continues to quote 24 days and book freight on that lane. Supply chain managers must manually identify every affected PO, calculate the new arrival dates, and manually expedite critical shipments via air freight—a chaotic, error-prone process that takes days.

The Agentic Mode Selector handles this autonomously in near real-time.

1. **Sense:** The Risk Agent, continuously monitoring global news APIs and maritime advisories, ingests alerts about the Red Sea closure.
2. **Plan:** When a new shipment request from Mumbai to Rotterdam enters the queue, the Orchestrator queries the Market Rate Agent. The API might still return the outdated 24-day transit time if the carrier hasn't updated their systems. However, the Risk Agent intervenes, flagging the lane and applying a semantic override: "Due to Red Sea rerouting, add 14 days to all Ocean transit times on this lane. Risk of further delay is High."
3. **Optimize:** The Orchestrator recalculates the utility. The new 38-day transit time violates the Required Delivery Date for a critical component. The agent evaluates alternatives: pure Air Freight (too expensive, exceeds budget threshold), or a Sea-Air multimodal route (Ocean to Dubai, Air to Frankfurt, Truck to Rotterdam).

4. **Act:** The agent calculates that the Sea-Air route meets the delivery date, minimizes the cost premium compared to pure air freight, and keeps the factory running. It selects this mode, drafts a detailed justification explaining the Red Sea impact, and routes it to the HITL dashboard for final execution.

1.51.2 Token Economics and Deployment Strategy

Deploying this level of intelligence requires careful management of token economics and API latency. A global manufacturer might process 10,000 shipments a day. Running a complex, multi-agent LangGraph workflow with a frontier model like Gemini 3.5 Pro for every single parcel is neither cost-effective nor necessary.

The best practice is a **Tiered Routing Architecture**.

- **Tier 1 (Deterministic):** 80% of shipments are standard, low-value, and low-urgency. These are processed by a traditional, deterministic rules engine (or a very small, fast model like Gemini 3.5 Flash-Lite) at a cost of fractions of a cent per transaction.
- **Tier 2 (Agentic Exception Management):** The deterministic engine includes trigger conditions: if the shipment value exceeds $50,000, if the required delivery date is within 7 days, or if the origin/destination matches a known active disruption zone (flagged by a background risk-monitoring agent), the shipment is kicked out of Tier 1 and routed to the Agentic Mode Selector.

By reserving the heavy cognitive lifting for exceptions and high-stakes shipments, organizations can achieve the resilience of an autonomous supply chain while maintaining a highly efficient cloud infrastructure footprint. The cost of running the agentic workflow (perhaps $0.05 to $0.15 per complex transaction in token costs) is instantly justified when the agent successfully avoids a single $50,000 line-down event or prevents the unnecessary booking of a $20,000 air charter.

Key Takeaways

- **Context Over Rules:** Static routing guides fail during disruptions. Agentic mode selectors succeed by dynamically engineering context from ERPs, MES, and unstructured intelligence before making a routing decision.
- **Multi-Objective Utility:** Freight selection is a trade-off between Cost, Time, Risk, and Carbon. Agents dynamically adjust the weights of these variables based on the specific urgency of the shipment (e.g., standard replenishment vs. line-down risk).
- **Separation of Concerns:** Do not use LLMs to guess freight rates. Architect systems where the LLM orchestrates deterministic API calls (to fetch spot rates) and applies its reasoning capabilities to risk assessment and trade-off analysis.
- **Tiered Deployment:** Optimize token costs and latency by using deterministic rules for standard shipments, routing only high-value, high-risk, or disrupted shipments to the heavy agentic reasoning workflow.
- **Sustainability as a Variable:** Agentic systems allow organizations to shift carbon tracking from a post-shipment reporting metric to a pre-shipment optimization

variable, autonomously selecting greener modes when time permits.

Learning Objectives

- Understand the architectural shift from static, dashboard-based risk scoring to continuous, agentic multi-modal signal ingestion.
- Design a Knowledge Graph-grounded reasoning system (GraphRAG) to calculate the N-tier "blast radius" of geopolitical, financial, or environmental events.
- Implement a multi-agent workflow that autonomously drafts and executes mitigation strategies (e.g., dual-sourcing, inventory buffering) before disruptions impact the manufacturing line.
- Evaluate the trade-offs between false-positive alert fatigue and autonomous action thresholds using a tiered agent architecture.

While the Agentic Mode Selector discussed in the previous section handles the dynamic routing of freight already in transit, true supply chain resilience requires looking much further upstream. The modern manufacturing supply chain is a highly optimized, deeply interconnected, and inherently fragile network. The pursuit of Just-In-Time (JIT) manufacturing and lean inventory has stripped away the buffers that historically absorbed shocks. Today, a localized disruption—a factory fire in Malaysia, a sudden export restriction in China, a cyberattack on a logistics provider in Germany, or a localized weather event—can cascade through the tiers of a supply chain, resulting in catastrophic line-down events for an Original Equipment Manufacturer (OEM) weeks later.

Traditional Supply Chain Risk Management (SCRM) systems have largely failed to prevent these cascading failures. They suffer from three fatal flaws: they are reactive, they lack N-tier visibility, and they are passive. Most SCRM platforms rely on static risk scores (e.g., financial health ratings updated quarterly) or keyword-based news scraping that floods procurement teams with thousands of irrelevant alerts (alert fatigue). Furthermore, when a genuine alert is generated, the system merely displays it on a dashboard. It requires a human to read the alert, manually cross-reference it against the Bill of Materials (BOM) in the ERP, identify the affected parts, and manually initiate mitigation. By the time this human-driven process is complete, the window for securing alternative supply has often closed.

The **Agentic Supplier Risk & Disruption Monitor** represents a paradigm shift from passive dashboards to active, autonomous resilience. By combining Large Language Models (LLMs) with Knowledge Graphs (GraphRAG) and multi-agent orchestration, we can build systems that continuously read global signals, autonomously calculate the exact blast radius of an event across multiple supply chain tiers, and proactively execute mitigation workflows before the disruption materializes at the factory gate.

1.52 The N-Tier Visibility Problem and GraphRAG

The most significant challenge in supply chain risk is the lack of visibility beyond Tier 1 (direct suppliers). An automotive OEM knows who supplies their infotainment head unit (Tier 1). They likely do not know who supplies the display panel to the head unit manufacturer (Tier 2), and they almost certainly do not know who supplies the specific display driver IC (integrated circuit) to the panel manufacturer (Tier 3).

When an earthquake strikes a specific industrial park in Taiwan, a keyword-based news alert might flag the event. But without N-tier visibility, the OEM cannot answer the critical question: *"Does this earthquake affect my production line?"*

To solve this, agentic systems rely on a **Supply Chain Knowledge Graph**. Unlike standard relational databases or vector stores, a Knowledge Graph models the supply chain topologically. Nodes represent entities (Companies, Facilities, Parts, Finished Goods), and edges represent relationships (`SUPPLIES`, `LOCATED_IN`, `PART_OF_BOM`).

When an agent utilizes **GraphRAG** (Graph Retrieval-Augmented Generation), it does not just perform a semantic search for similar text. It executes graph traversals. If the agent ingests a news article about a fire at "Acme Semiconductor Fab 2 in Hsinchu," it extracts the entity (Acme Semiconductor) and the location (Hsinchu). It then queries the Knowledge Graph:

1. Find all facilities matching "Acme Semiconductor" in "Hsinchu".
2. Traverse outgoing `PRODUCES` edges to find all Parts manufactured there.
3. Traverse incoming `PURCHASES` edges to find all Tier 2 suppliers buying those Parts.
4. Recursively traverse `PART_OF_BOM` edges upward until reaching the OEM's Finished Goods.

The result of this traversal is the exact "Blast Radius"—a deterministic list of the OEM's specific products and revenue streams at risk. The LLM is then used to synthesize this complex graph data into a human-readable impact assessment and to formulate a mitigation plan based on the specific constraints of the affected parts.

1.53 Architecture of the Agentic Risk Monitor

Building this system requires a multi-agent architecture that separates the high-volume, low-complexity task of signal ingestion from the low-volume, high-complexity task of mitigation planning.

The architecture consists of three primary layers:

1.53.1 1. The Signal Ingestion & Triage Layer

The world generates millions of data points daily. Feeding all of this into a frontier model like Gemini 3.5 Pro would be prohibitively expensive and slow. Instead, we deploy a fleet of lightweight **Triage Agents** (often powered by Gemini 3.5 Flash-Lite or similar fast, cost-effective models).

These agents monitor APIs such as GDELT (Global Database of Events, Language, and Tone), NOAA weather feeds, maritime AIS data, and financial SEC filings. Their sole job is binary classification and entity extraction: *"Is this piece of unstructured data a supply chain disruption event? If yes, what are the entities involved?"*

If a Triage Agent reads a news article about a CEO resigning, it classifies it as low supply chain risk and discards it. If it reads an article about a dockworker strike in Long Beach, it flags it as high risk, extracts the location entity ("Port of Long Beach"), and passes the structured payload to the next layer. This triage mechanism solves the alert fatigue problem that plagues traditional SCRM systems.

1.53.2 2. The Graph Reasoning Layer

Once a verified threat is identified, the payload is handed to the **Reasoning Agent** (powered by a frontier model capable of complex logic and tool use). This agent is equipped with tools to query the Supply Chain Knowledge Graph (e.g., generating Cypher queries for Neo4j) and the enterprise ERP (e.g., SAP or Oracle via API).

The Reasoning Agent executes the GraphRAG traversal to determine the blast radius. Crucially, it then queries the ERP to check current inventory levels. A disruption at a Tier 2 supplier is only a crisis if the OEM has no safety stock. If the agent discovers that the OEM holds 90 days of inventory for the affected component, and the disruption is estimated to last 14 days, the agent downgrades the severity of the alert and simply logs it for monitoring.

1.53.3 3. The Mitigation & Action Layer

If the Reasoning Agent determines that the disruption will cause a stockout, it transitions from analysis to action. It queries the ERP for alternative approved suppliers for the affected component. It checks spot market APIs for availability and pricing. It then formulates a mitigation plan.

Instead of just sending an email, the agent uses its API tools to actively draft a Purchase Order (PO) for the alternative supplier, or it updates the MRP (Material Requirements Planning) parameters in the ERP to temporarily increase safety stock levels for related components. It then routes this proposed action to a human procurement manager for a single-click approval (Human-in-the-Loop).

1.54 Implementation: Building the Risk Reasoning Workflow

To demonstrate this, we will implement the core logic of the Reasoning Agent using LangGraph. This workflow takes a triaged risk event, queries a mocked Knowledge Graph to find the blast radius, checks inventory, and proposes a mitigation strategy.

Listing 1.35: LangGraph implementation of the Supplier Risk Reasoning Workflow

```python
import operator
from typing import TypedDict, Annotated, List, Dict, Any
from langchain_google_vertexai import ChatVertexAI
from langchain_core.messages import SystemMessage, HumanMessage
from langgraph.graph import StateGraph, END

# 1. Define the State
class RiskState(TypedDict):
    event_description: str
    extracted_entities: List[str]
    blast_radius: List[Dict[str, Any]]
    inventory_status: Dict[str, Any]
    mitigation_plan: str
    requires_action: bool

llm = ChatVertexAI(model_name="gemini-3.5-pro", temperature=0.1)

# 2. Node: GraphRAG Traversal (Mocked Knowledge Graph Query)
```

```
def calculate_blast_radius(state: RiskState):
    """Simulates querying a Neo4j Knowledge Graph to find affected BOMs."""
    entities = state["extracted_entities"]
    # In production, the LLM would generate a Cypher query here based on the entities.
    # For demonstration, we mock the graph response.

    print(f"--- Traversing Graph for entities: {entities} ---")

    # Mocked Graph Response: The event affects a Tier 2 supplier of Microcontrollers
    blast_radius = [
        {
            "tier_2_supplier": "Taiwan Semi Fab 4",
            "affected_part": "MCU-8922",
            "tier_1_assembler": "Bosch Electronics",
            "oem_finished_good": "Model X Infotainment System",
            "revenue_at_risk_usd": 4500000
        }
    ]
    return {"blast_radius": blast_radius}

# 3. Node: ERP Inventory Check
def check_inventory(state: RiskState):
    """Simulates querying SAP/ERP for current stock levels of affected parts."""
    affected_parts = [item["affected_part"] for item in state["blast_radius"]]

    print(f"--- Checking ERP Inventory for: {affected_parts} ---")

    # Mocked ERP Response
    inventory_status = {
        "MCU-8922": {
            "current_stock": 1200,
            "daily_burn_rate": 100,
            "days_of_cover": 12,
            "alternative_suppliers": ["NXP (Approved)", "Texas Instruments (Pending)"]
        }
    }
    return {"inventory_status": inventory_status}

# 4. Node: Mitigation Strategy Generation
def generate_mitigation(state: RiskState):
    """Uses the LLM to analyze the blast radius and inventory to propose action."""
    prompt = f"""
    You are an expert Supply Chain Risk Agent. Analyze the following disruption:
    Event: {state['event_description']}

    Blast Radius (Affected BOMs):
    {state['blast_radius']}

```

```
    Current Inventory Status:
    {state['inventory_status']}

    Determine if immediate action is required. If Days of Cover < 30, action
    is REQUIRED.
    Draft a specific mitigation plan (e.g., spot buy, dual-sourcing).
    """

    structured_llm = llm.with_structured_output(
        schema={
            "type": "object",
            "properties": {
                "requires_action": {"type": "boolean"},
                "mitigation_plan": {"type": "string"}
            }
        }
    )

    decision = structured_llm.invoke([
        SystemMessage(content="You are an autonomous supply chain orchestrator.
"),
        HumanMessage(content=prompt)
    ])

    return {
        "requires_action": decision["requires_action"],
        "mitigation_plan": decision["mitigation_plan"]
    }

# 5. Compile the Graph
workflow = StateGraph(RiskState)
workflow.add_node("GraphTraversal", calculate_blast_radius)
workflow.add_node("InventoryCheck", check_inventory)
workflow.add_node("MitigationPlanner", generate_mitigation)

workflow.set_entry_point("GraphTraversal")
workflow.add_edge("GraphTraversal", "InventoryCheck")
workflow.add_edge("InventoryCheck", "MitigationPlanner")
workflow.add_edge("MitigationPlanner", END)

risk_agent_app = workflow.compile()

# Example Execution
initial_state = {
    "event_description": "A magnitude 6.5 earthquake struck Hsinchu, Taiwan,
    causing power outages at major semiconductor fabrication plants.",
    "extracted_entities": ["Hsinchu", "Taiwan", "semiconductor"]
}

# result = risk_agent_app.invoke(initial_state)
# print(result["mitigation_plan"])
```

In a production environment, the `calculate_blast_radius` node utilizes a tool that allows the LLM to dynamically write and execute Cypher queries against a Neo4j database. This allows the agent to handle novel, unforeseen queries that hardcoded SQL statements cannot accommodate.

1.55 Maturity Curve: From Reactive to Autonomous

The implementation of an Agentic Risk Monitor represents the final stage in the evolution of supply chain risk management. Organizations typically progress through four distinct stages of maturity.

1. **Reactive (Manual):** The organization learns about a disruption when a supplier calls to say a shipment will be late. Mitigation involves expensive, panicked air-freight expediting and manual spreadsheet analysis.
2. **Predictive (Dashboards):** The organization purchases a risk monitoring SaaS tool. They receive alerts about global events, but lack N-tier visibility. Procurement teams suffer from alert fatigue, ignoring the dashboard until a crisis occurs.
3. **Prescriptive (AI Assist):** LLMs are introduced to summarize news alerts and suggest potential actions. However, the AI is disconnected from the ERP and Knowledge Graph. It acts as an advisor, but humans must still do the heavy lifting of data correlation and execution.
4. **Autonomous (Agentic):** The system described in this chapter. The AI is integrated with the Knowledge Graph and ERP. It senses the event, calculates the exact blast radius, checks inventory constraints, and autonomously drafts the PO for the alternative supplier, requiring only a single human click to execute.

1.56 Managing False Positives and The Action Space

A critical design consideration for the Agentic Risk Monitor is defining the "Action Space"—what the agent is actually allowed to do without human intervention. If an agent hallucinates a risk or overreacts to a minor news story, autonomously shifting millions of dollars of procurement to a secondary supplier could be disastrous.

To manage this, we utilize a **Risk Triage Matrix** to define the boundaries of autonomy based on Probability and Impact.

For low-impact events, the agent is granted full autonomy to make minor adjustments, such as slightly increasing the safety stock parameters in the MRP system (Automated Buffer). For high-impact events (e.g., a major geopolitical shift or a Tier 1 bankruptcy), the agent is restricted to a "Draft and Propose" action space. It prepares the contingency plan, drafts the necessary communications and POs, but halts execution at a Human-in-the-Loop (HITL) gateway.

Furthermore, to combat false positives generated by unreliable news sources, robust agentic systems employ a **Critic Agent** pattern. Before the Reasoning Agent is allowed to trigger a mitigation workflow, a separate Critic Agent is invoked to verify the source credibility. The Critic Agent cross-references the initial alert against multiple independent data streams (e.g., verifying a news report of a port strike against live AIS vessel tracking data to see if ships are actually queuing outside the port). Only if the Critic Agent achieves a high confidence consensus is the workflow allowed to proceed.

By combining N-tier Knowledge Graphs, multi-modal signal ingestion, and carefully bounded autonomous action spaces, the Agentic Supplier Risk Monitor transforms supply chain resilience from a reactive firefighting exercise into a proactive, continuous optimization loop.

Key Takeaways

- **N-Tier Visibility is Mandatory:** Traditional risk monitoring fails because it only sees Tier 1 suppliers. Agentic systems require a Supply Chain Knowledge Graph to trace the topological "blast radius" of disruptions deep within the sub-tiers.
- **GraphRAG over Vector Search:** For supply chain risk, semantic similarity (Vector Search) is insufficient. Agents must use GraphRAG to traverse relationships (`SUPPLIES`, `PART_OF_BOM`) to deterministically link global events to specific finished goods.
- **Tiered Agent Architecture:** To manage token costs and latency, use small, fast models (Gemini Flash) for continuous signal triage, and reserve frontier models (Gemini Pro) for complex blast-radius reasoning and mitigation planning.
- **Contextualize with ERP Data:** A disruption is only a crisis if inventory is low. The Reasoning Agent must always query the ERP for current safety stock and days-of-cover before triggering a mitigation workflow.
- **Bounded Autonomy:** Define the agent's action space using a Risk Triage Matrix. Allow autonomous execution for low-impact buffering, but enforce Human-in-the-Loop (HITL) gateways for high-impact, high-cost mitigation strategies like dual-sourcing.

Learning Objectives

- Understand the limitations of traditional Transportation Management Systems (TMS) in addressing Scope 3 emissions and dynamic, multi-objective optimization.
- Architect a multi-agent system capable of ingesting real-time telematics, weather, and carrier data to calculate predictive carbon footprints using the GLEC framework.
- Implement a dynamic utility function and Pareto optimization logic within an LLM-driven reasoning loop to balance cost, delivery SLAs, and CO2 emissions.
- Design Agent-to-Agent (A2A) negotiation protocols for autonomous freight bidding and exception management in the event of supply chain disruptions.

1.57 The Scope 3 Challenge and the Limits of Static Routing

The manufacturing and logistics sectors are undergoing a profound paradigm shift, driven not merely by the desire for operational efficiency, but by stringent, globally enforced sustainability mandates. Regulatory frameworks such as the European Union's Corporate Sustainability Reporting Directive (CSRD), the California Climate Corporate Data Accountability Act (SB

253), and the SEC's evolving climate disclosure rules have transformed carbon emissions from a secondary public relations concern into a primary, auditable financial liability. For most manufacturing enterprises, the vast majority of these emissions—often exceeding 80% of their total carbon footprint—fall under Scope 3: indirect emissions that occur in the upstream and downstream value chain. Within Scope 3, transportation and distribution represent one of the largest, most volatile, and historically least controllable categories.

The fundamental challenge in decarbonizing logistics lies in the physics of freight and the rigid architecture of legacy software. The carbon footprint of a shipment is a complex, non-linear function of weight, distance, transportation mode (air, ocean, rail, road), vehicle efficiency, fuel type (diesel, LNG, electric, SAF), and real-world operational conditions such as idling time, empty miles, and weather-induced detours. Traditional Transportation Management Systems (TMS) were architected in an era where the sole objectives were minimizing cost and meeting delivery Service Level Agreements (SLAs). They operate on static routing guides, historical rate cards, and deterministic algorithms (like the classic Vehicle Routing Problem solvers) that treat the supply chain as a closed, predictable system.

When a traditional TMS attempts to incorporate carbon emissions, it typically does so as a post-facto reporting exercise. It takes the completed shipment data, multiplies the distance by a generic, static emission factor (e.g., average grams of CO2 per ton-kilometer for a generic diesel truck), and outputs a retrospective report. This is fundamentally inadequate for active decarbonization. It is the equivalent of driving a car by only looking in the rearview mirror. To actually reduce emissions, the optimization must be predictive, dynamic, and integrated directly into the procurement and routing decision-making process before the freight is ever tendered.

This introduces the "Logistics Trilemma": the inherent tension between Cost, Speed, and Carbon. The fastest route (air freight) is the most carbon-intensive and expensive. The greenest route (ocean + rail) is the slowest. The cheapest route might involve consolidating loads, which delays departure times. Resolving this trilemma requires a system capable of multi-objective optimization. However, the optimal balance is not static; it changes daily based on external context. If a critical manufacturing line is about to shut down due to a missing component, the SLA becomes paramount, and the system must authorize the carbon-intensive air freight. If the inventory buffer is healthy, the system should autonomously default to the slower, greener, cheaper rail option.

Legacy systems fail here because they lack semantic understanding and contextual awareness. A standard TMS cannot read an email from a supplier indicating a production delay, cross-reference it with live port congestion data in Rotterdam, check the ERP for current safety stock levels, and autonomously decide that a hybrid sea-air routing is the only way to meet the SLA without exceeding the quarterly carbon budget.

This is where Agentic AI transforms the paradigm. A CO2-Aware Routing Optimizer is not merely a better algorithm; it is a cognitive architecture. It utilizes Large Language Models (LLMs) to process unstructured context (emails, news, weather alerts), integrates with Knowledge Graphs to understand the topological dependencies of the supply chain, and employs specialized tools to calculate precise, predictive emissions using industry standards like the Global Logistics Emissions Council (GLEC) framework. By framing routing as an autonomous, goal-seeking agent workflow, we move from static, cost-only execution to dynamic, context-aware, multi-objective orchestration. The agent does not just find a route; it negotiates the best possible reality given the constraints of the physical world.

1.58 Architecture of the CO2-Aware Routing Agent

Building a production-grade CO2-Aware Routing Optimizer requires a robust, multi-agent architecture that bridges the gap between high-level semantic reasoning and low-level deterministic calculations. The system cannot rely solely on an LLM to "guess" the best route or calculate emissions, as LLMs are notoriously poor at precise arithmetic and spatial graph traversal. Instead, the architecture follows the Planner-Executor-Evaluator pattern, where the LLM acts as the orchestrator, delegating specific tasks to deterministic APIs, graph databases, and specialized sub-agents.

The architecture is built upon four primary pillars: Data Ingestion & Contextualization, the Emissions Calculation Engine, the Graph Search & Routing Engine, and the Agentic Orchestrator.

1.58.1 Data Ingestion and Contextualization

The foundation of the system is real-time visibility. The Ingestion Agent continuously monitors multiple data streams. From the ERP (e.g., SAP S/4HANA), it extracts the demand signal: what needs to be shipped, its weight, volume, origin, destination, and the required delivery date (SLA). From supply chain visibility platforms (like Project44 or FourKites), it ingests real-time telematics: the current location of vessels, port congestion indices, and border crossing wait times. Furthermore, it ingests unstructured data, such as weather forecasts and geopolitical news alerts. The Ingestion Agent's primary role is to synthesize this disparate data into a unified state object—a "Routing Request Context"—that the downstream agents can understand.

1.58.2 The Emissions Calculation Engine (GLEC Compliant)

To ensure regulatory compliance and auditability, the system must calculate emissions using standardized methodologies, specifically the GLEC Framework. The Emissions Calculator is a deterministic tool (often a Python microservice wrapping an API like Searoutes or the Smart Freight Centre's tools) that the LLM can invoke. It calculates emissions on a Well-to-Wheel (WTW) basis, accounting for both the fuel combustion (Tank-to-Wheel) and the upstream fuel production (Well-to-Tank).

When the LLM proposes a route, it passes the parameters (Origin, Destination, Mode, Weight, Carrier/Vessel specific data if available) to this tool. The tool returns the predicted CO2 equivalent (CO2e) in kilograms. Crucially, because this is an agentic system, it can handle data sparsity. If the exact vessel is unknown, the LLM can instruct the tool to use trade-lane averages. If the carrier provides primary data via an API, the agent prioritizes that over generic emission factors, ensuring the highest possible data quality for Scope 3 reporting.

1.58.3 Graph Search and Routing Engine

Supply chain networks are inherently topological. Finding the physical paths between Point A and Point B is a graph traversal problem, not a language generation problem. The architecture utilizes a Graph Database (like Neo4j) or a specialized routing engine (like OSRM for road, or specialized maritime routing APIs) to generate the physical route options. The LLM does not invent the routes; it queries the Graph Engine. For example, the LLM might query: *"Return the top 3 multimodal routes from Shenzhen to Munich, prioritizing one for speed, one for lowest cost, and one for lowest carbon."* The Graph Engine executes the A* or Dijkstra algorithms and returns the deterministic paths, distances, and transit times.

1.58.4 The Agentic Orchestrator

This is the cognitive center of the system, typically implemented using a stateful framework like LangGraph or the Google Agent Development Kit (ADK). The Orchestrator receives the Routing Request Context, queries the Graph Engine for physical options, and then iteratively queries the Emissions Calculator and Carrier Rate APIs for each option.

Once it has assembled a matrix of options (each with a Cost, Time, and CO2 value), the Orchestrator applies a dynamic utility function to make the final decision. This is where the LLM's reasoning capabilities shine. It can interpret soft constraints and business rules. For instance, if the ERP context indicates that the shipment contains hazardous materials, the Orchestrator knows to filter out certain carriers or routes. If the context indicates a high risk of a port strike at the primary destination, the Orchestrator can autonomously penalize that route's utility score, favoring a slightly more expensive but more reliable alternative.

This architecture ensures that the LLM is used for what it does best—semantic reasoning, context synthesis, and complex decision-making—while relying on deterministic systems for math, spatial routing, and regulatory compliance. The result is a system that is both highly adaptable and strictly auditable.

1.59 Dynamic Trade-off Optimization: Cost, Time, and Carbon

The core intelligence of the CO2-Aware Routing Optimizer lies in its ability to evaluate competing routing options and select the optimal path. This is not a simple minimization problem; it is a multi-objective optimization challenge where the variables—Cost, Time, and Carbon—are often inversely correlated. To solve this, the agentic system employs a dynamic utility function that calculates the "Pareto efficiency" of each route, adjusted in real-time by an Internal Carbon Pricing (ICP) mechanism and SLA penalty functions.

1.59.1 The Mathematics of the Utility Function

To allow the LLM to make objective, mathematically sound decisions, we must normalize the three variables into a single scalar value: the Total Utility Cost (U_{total}). The agent seeks to minimize this value.

The function can be expressed as:

$$U_{total} = C_{freight} + (E_{co2} \times P_{carbon}) + P_{sla}(T_{transit}, T_{required})$$

Where:

- $C_{freight}$ is the hard financial cost of the freight quote in USD.
- E_{co2} is the predicted emissions in metric tons of CO2e (from the GLEC tool).

- P_{carbon} is the Internal Carbon Price in USD per ton. This is a critical lever. By setting a high ICP (e.g., $150/ton), the business mathematically forces the agent to favor greener routes, even if the hard freight cost is higher.
- P_{sla} is a penalty function applied if the transit time ($T_{transit}$) exceeds the required delivery date ($T_{required}$). This is not a linear function; it is typically an exponential step function. If the delivery is early or on time, the penalty is zero. If it is one day late, the penalty might be $500. If it is five days late and causes a factory line stoppage, the penalty approaches infinity, effectively vetoing the route.

1.59.2 Implementing the Optimization Loop in LangGraph

In a production environment, this logic is encapsulated within a specific node in the agent's cognitive graph. The LLM is provided with the raw data for the available routes and the current business parameters (ICP, SLA criticality). It is instructed to calculate the utility score for each route and output a structured JSON decision.

Below is a simplified implementation of this reasoning node using Python and LangChain/LangGraph, demonstrating how the agent evaluates three distinct multi-modal options.

Listing 1.36: LangGraph Node for Multi-Objective Route Optimization

```python
import json
from typing import List, Dict, Any
from langchain_core.messages import SystemMessage, HumanMessage
from langchain_google_genai import ChatGoogleGenerativeAI
from pydantic import BaseModel, Field

# Define the structured output schema for the LLM
class RouteDecision(BaseModel):
    selected_route_id: str = Field(description="The ID of the chosen route.")
    reasoning: str = Field(description="Detailed explanation of the trade-offs
     considered.")
    calculated_utility_scores: Dict[str, float] = Field(description="The
    U_total for each route.")

# Initialize the reasoning model (Gemini 3.5 Pro for complex logic)
llm = ChatGoogleGenerativeAI(model="gemini-3.5-pro", temperature=0.1)

def optimize_route_node(state: dict) -> dict:
    """
    Evaluates available routes based on Cost, Time, and Carbon.
    """
    shipment_context = state["shipment_context"]
    available_routes = state["available_routes"]

    # Business Parameters (In production, fetched dynamically from a policy
    store)
    internal_carbon_price = 100.0  # USD per metric ton of CO2e
    sla_penalty_per_day = 2000.0   # USD penalty per day late

    system_prompt = """
```

```
    You are an autonomous Supply Chain Routing Agent. Your goal is to select the optimal
    freight route by minimizing the Total Utility Cost (U_total).

    Formula: U_total = Freight_Cost + (CO2_Emissions_Tons * Internal_Carbon_Price) + SLA_Penalty

    Business Parameters:
    - Internal Carbon Price: $100 per metric ton
    - SLA Penalty: $2000 per day late (0 if on time or early)

    Analyze the provided routes, calculate the U_total for each, and select the lowest score.
    If two routes have similar scores (within 5%), prefer the route with lower CO2 emissions.
    """

    human_prompt = f"""
    Shipment Context:
    - Required Delivery Date: {shipment_context['required_date']}
    - Criticality: {shipment_context['criticality']}

    Available Routes:
    {json.dumps(available_routes, indent=2)}
    """

    # Bind the schema to force structured JSON output
    structured_llm = llm.with_structured_output(RouteDecision)

    # Invoke the model
    decision = structured_llm.invoke([
        SystemMessage(content=system_prompt),
        HumanMessage(content=human_prompt)
    ])

    print(f"Agent Selected Route: {decision.selected_route_id}")
    print(f"Reasoning: {decision.reasoning}")

    # Update state and pass to the execution node
    return {"selected_route": decision.selected_route_id, "routing_decision": decision.dict()}

# Example State Data
mock_state = {
    "shipment_context": {
        "required_date": "2026-06-28",
        "criticality": "High"
    },
    "available_routes": [
        {
            "route_id": "R1_AIR",
            "mode": "Air Freight",
            "freight_cost_usd": 8500,
            "co2_emissions_tons": 12.5,
```

```
            "estimated_arrival": "2026-06-25" # 3 days early
        },
        {
            "route_id": "R2_OCEAN_RAIL",
            "mode": "Ocean + Rail",
            "freight_cost_usd": 2200,
            "co2_emissions_tons": 1.8,
            "estimated_arrival": "2026-06-30" # 2 days late
        },
        {
            "route_id": "R3_TRUCK_TEAM",
            "mode": "Team Transit Truck",
            "freight_cost_usd": 4100,
            "co2_emissions_tons": 4.2,
            "estimated_arrival": "2026-06-27" # 1 day early
        }
    ]
}

# optimize_route_node(mock_state)
```

In the mock scenario above, the agent evaluates the options. Route 1 (Air) is fast but has a massive carbon penalty ($12.5 \times 100 = \$1250$), making its utility score $\$9750$. Route 2 (Ocean) is cheap and green, but it arrives two days late, incurring a massive SLA penalty ($2 \times 2000 = \$4000$), making its utility score $\$6380$. Route 3 (Truck) arrives on time (no penalty), has moderate cost ($4100), and moderate carbon penalty ($420). Its utility score is $\$4520$. The agent autonomously selects Route 3, perfectly balancing the trilemma without human intervention.

1.60 Autonomous Carrier Negotiation and Exception Management

Selecting the optimal route in a simulation is only half the battle; the agent must execute this decision in the real world. This introduces the complexities of carrier negotiation, capacity constraints, and exception management. In a fully mature agentic ecosystem, this is achieved through Agent-to-Agent (A2A) protocols, where the Shipper Agent (our Routing Optimizer) negotiates directly with Carrier Agents.

1.60.1 Agent-to-Agent (A2A) Freight Bidding

Legacy freight procurement relies on static EDI (Electronic Data Interchange) connections or human-operated load boards. Agentic routing utilizes dynamic API integrations and open protocols like the Agent Payments Protocol (AP2) to conduct micro-auctions in real-time.

When our Routing Optimizer selects "Route 3 (Truck)" from the previous example, it does not just send a blind tender. It broadcasts a structured request to a network of approved Carrier Agents. This request includes the origin, destination, weight, required SLA, and crucially, a request for the carrier's specific emission factors.

A Carrier Agent representing a logistics company with a fleet of Electric Vehicles (EVs) might respond with a slightly higher freight cost but a near-zero CO2 emission profile. A competing Carrier Agent with older diesel trucks might offer a lower price but higher emissions. Because our Shipper Agent uses the dynamic utility function (incorporating the Internal Carbon Price), it can mathematically justify awarding the load to the EV carrier, autonomously executing the company's sustainability strategy at the transactional level.

To prevent "greenwashing," the A2A protocol requires verifiable credentials. The Carrier Agent must provide cryptographic proof of their emission factors (e.g., verified telemetry data from their fleet, audited by a third party), which the Shipper Agent's Critic Node verifies before accepting the bid.

1.60.2 Dynamic Exception Management and Re-routing

The physical supply chain is chaotic. Trucks break down, ports strike, and weather systems delay vessels. A static TMS fails catastrophically during exceptions, requiring human planners to manually expedite freight, usually defaulting to carbon-intensive air freight in a panic.

The Agentic Routing Optimizer handles exceptions continuously. Because it is integrated with real-time telematics and weather APIs, it operates in a continuous Sense-Plan-Act loop.

Consider a scenario where a shipment is currently on a rail car heading to a port, but the Ingestion Agent detects a sudden labor strike at that specific port. 1. **Sense:** The agent detects the anomaly via a news API and cross-references it with the shipment's current GPS coordinates. 2. **Plan:** The agent calculates the "blast radius" of the delay. It realizes the shipment will miss its ocean vessel connection, violating the SLA. It immediately queries the Graph Engine for alternative routes from the train's current location. It evaluates offloading the container to a truck and driving to an alternative port, versus flying the cargo. 3. **Act:** It runs the utility function. It determines that trucking to an alternative port preserves the SLA while emitting 80% less CO2 than air freight. It autonomously contacts a Carrier Agent, secures a truck, issues the re-routing instructions to the rail operator, and updates the ERP—all within seconds of the news breaking.

This level of autonomous, continuous optimization is impossible with traditional software. By embedding carbon accounting directly into the reasoning loop of an autonomous agent, organizations can finally decouple supply chain resilience from carbon intensity, achieving both operational excellence and verifiable sustainability.

Key Takeaways

- **Carbon as a Primary Variable:** Agentic routing elevates CO2 emissions from a post-facto reporting metric to a primary, dynamic variable in the routing decision process, alongside cost and time.
- **Dynamic Utility Functions:** Use LLMs to evaluate a mathematical utility function that normalizes Cost, SLA penalties, and Carbon (via an Internal Carbon Price) to find the Pareto optimal route.
- **Deterministic Delegation:** LLMs should orchestrate the workflow, but must delegate spatial routing to Graph Engines and emissions math to GLEC-compliant deterministic calculators to ensure accuracy and auditability.

Quality Assurance & Inspection

Figure 1.31: The transition from reactive inspection to autonomous, agentic quality assurance.

Learning Objectives

- **Understand the paradigm shift** from static, rule-based quality control to autonomous, agentic quality assurance that reasons over multi-modal data.
- **Identify core architectural patterns** for integrating specialized agents (vision, time-series, text) into the end-to-end manufacturing quality lifecycle.
- **Design closed-loop systems** where agents not only detect defects but autonomously orchestrate root-cause analysis, traceability, and corrective actions (CAPA).
- **Evaluate the role of agentic systems** in maintaining continuous compliance, generating digital documentation, and ensuring audit readiness in regulated environments.

For decades, industrial quality assurance (QA) has operated as a reactive, siloed, and highly manual discipline. Even as Industry 4.0 introduced advanced sensors and computer vision to the shop floor, the fundamental architecture of quality management remained fragmented. A high-speed camera might successfully flag a microscopic scratch on a machined part, but the subsequent steps—halting the line, querying the Manufacturing Execution System (MES) for the batch genealogy, analyzing the Statistical Process Control (SPC) charts for tool wear, and drafting a Corrective and Preventive Action (CAPA) report in the Quality Management System (QMS)—still require intense human intervention. This disconnect between isolated detection and systemic resolution is the primary driver of the Cost of Poor Quality (COPQ), which routinely consumes 10% to 15% of total sales revenue in complex manufacturing sectors. The bottleneck is no longer our ability to see defects; it is our capacity to reason about them, contextualize them, and act upon them at machine speed.

Agentic Artificial Intelligence fundamentally rewrites this equation by bridging the gap between perception and orchestration. In this chapter, we define *Agentic Quality Assurance* as the deployment of autonomous, goal-seeking systems capable of managing the entire quality lifecycle. Unlike traditional Automated Quality Control (AQC)—which relies on rigid, deterministic thresholds to execute binary pass/fail decisions—an agentic system possesses semantic understanding and operational agency. When an agentic vision system detects an anomaly, it does not merely increment a defect counter. It dynamically invokes tools to pull the last hour of telemetry from the programmable logic controller (PLC), cross-references the raw material lot numbers via the Enterprise Resource Planning (ERP) API, and synthesizes a probabilistic root-cause hypothesis. If the confidence threshold is met, the agent can autonomously quarantine the affected WIP (Work in Progress), adjust the upstream machine parameters to prevent recurrence, and pre-populate the regulatory compliance documentation for human review.

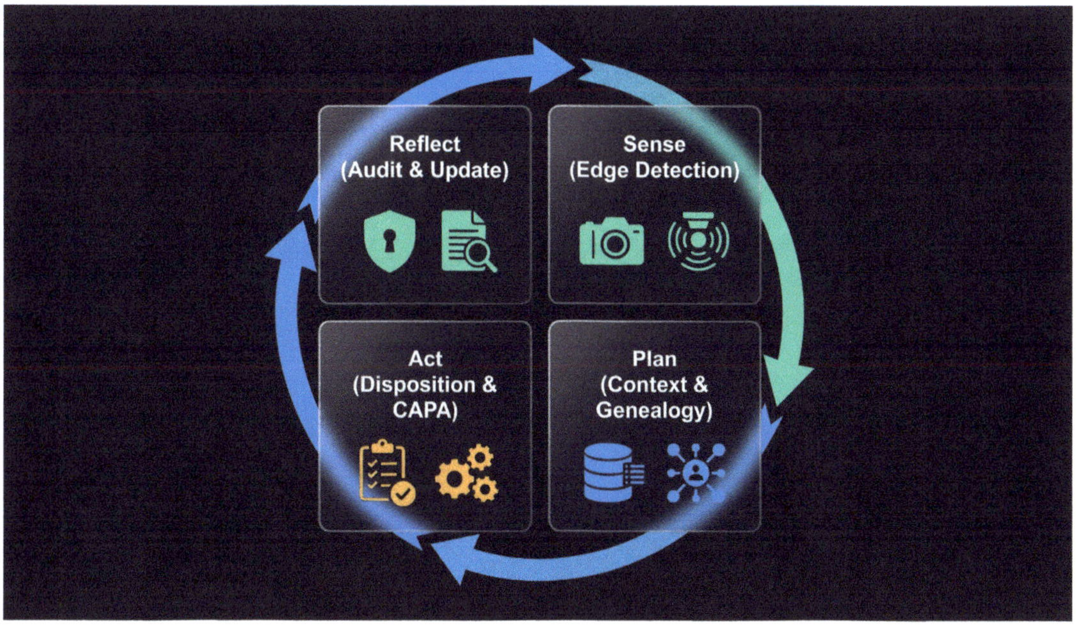

Figure 1.32: The Agentic Quality Assurance Lifecycle, moving from isolated edge detection to continuous, closed-loop enterprise reflection.

Architecting these systems requires a departure from monolithic AI models in favor of a multi-agent mesh. The modern QA architecture is a collaborative ecosystem of specialized actors. At the edge, low-latency Vision Defect Detectors and SPC Anomaly Agents operate on real-time data streams, utilizing lightweight models optimized for inference speed. These edge agents act as the sensory organs of the system. When they detect a deviation, they escalate the event to a centralized orchestration layer—often built on frameworks like LangGraph or the Google Agent Development Kit (ADK). Here, reasoning agents take over. A Traceability Copilot traverses the manufacturing graph to identify the blast radius of the defect, while a Digital COA (Certificate of Analysis) Reviewer ensures that supplier materials meet specifications. These agents communicate via standardized protocols, such as the Model Context Protocol (MCP), allowing them to seamlessly query legacy SQL databases, unstructured PDF repositories, and real-time MQTT brokers without requiring hardcoded integration logic for every new data source.

The business value of this rigorous, agentic approach extends far beyond simple scrap reduction; it fundamentally alters the compliance and audit posture of the enterprise. In highly regulated industries—such as pharmaceuticals (governed by FDA 21 CFR Part 11), aerospace (AS9100), and automotive (IATF 16949)—the burden of proof is as critical as the quality of the product itself. Traditional compliance relies on retrospective sampling and manual paperwork, leaving organizations vulnerable to audit findings due to human error or missing documentation. Agentic QA enables a shift to 100% inline inspection coupled with autonomous, real-time documentation. Every decision made by the agentic system is logged, cited, and cryptographically hashed, creating an immutable audit trail. By enforcing deterministic execution paths and strict Human-in-the-Loop (HITL) guardrails for critical dispositioning events, organizations can deploy autonomous systems that satisfy the most stringent regulatory requirements while drastically reducing the administrative overhead of quality management.

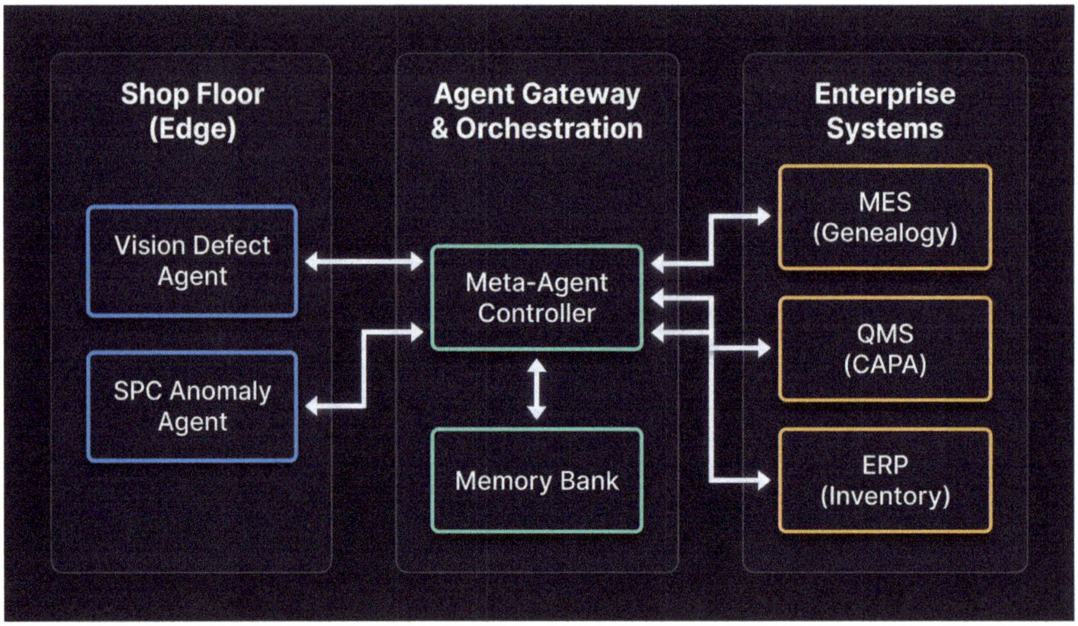

Figure 1.33: Multi-Agent QMS Architecture, illustrating the bidirectional flow of context and commands between edge sensors and enterprise systems.

What This Chapter Covers

To move from the theory of agentic quality to production-ready deployment, we must deconstruct the QA lifecycle into specific, implementable design patterns. The remainder of this chapter provides deep, technical blueprints for the six core agents that comprise a modern, autonomous Quality Management System. We will explore the architectural decisions, prompt engineering strategies, tool integrations, and failure modes for each of the following systems:

- **Vision Defect Detector:** Architecting edge-deployed multimodal agents that not only identify physical defects but generate semantic descriptions and severity classifications to inform downstream routing.
- **SPC Anomaly Agent:** Designing time-series reasoning agents that monitor Statistical Process Control charts, differentiate between common-cause and special-cause variation, and autonomously query machine telemetry for root-cause indicators.
- **Traceability & Genealogy Copilot:** Implementing graph-traversal agents capable of navigating complex MES databases to instantly map the forward and backward genealogy of a suspect component, isolating the exact blast radius of a quality event.
- **Digital COA/PPAP Reviewer:** Building document-understanding agents that ingest, parse, and validate complex supplier documentation (Certificates of Analysis and Production Part Approval Processes) against internal engineering specifications.
- **Complaint–CAPA Loop Orchestrator:** Deploying a multi-agent workflow that ingests customer complaints, correlates them with internal production data, and autonomously drafts comprehensive Corrective and Preventive Action plans for engineering review.

- **Audit Prep Summarizer:** Creating a specialized retrieval-augmented generation (RAG) agent that continuously compiles compliance evidence, maps internal processes to ISO/FDA standards, and generates natural-language audit defense packages.

By mastering these patterns, AI architects and quality engineers can transform their manufacturing operations from a state of reactive firefighting into a proactive, self-correcting, and continuously compliant ecosystem.

1.61 Vision Defect Detector

Learning Objectives

- **Deconstruct the semantic gap** in traditional Automated Optical Inspection (AOI) and understand how multimodal LLMs provide contextual reasoning for defect classification.
- **Architect a cascade inference pipeline** that balances the strict latency requirements of inline manufacturing (milliseconds) with the deep reasoning capabilities of agentic models.
- **Implement tool-calling vision agents** capable of autonomously querying engineering specifications (RAG) and executing disposition commands in the Manufacturing Execution System (MES).
- **Design robust prompt structures** for multimodal models to ensure deterministic, structured JSON outputs suitable for programmatic routing.

The front line of any industrial quality assurance system is perception. For decades, this perception has been governed by Automated Optical Inspection (AOI) systems. Traditional AOI relies on deterministic, rule-based algorithms—edge detection, blob analysis, and template matching—to identify deviations from a known good state. In recent years, these systems have been augmented by deep learning models, specifically Convolutional Neural Networks (CNNs) and Vision Transformers (ViTs), which excel at drawing bounding boxes around anomalies and assigning confidence scores. However, despite these advancements, the fundamental output of these systems remains a binary classification: Pass or Fail.

This binary paradigm creates a massive "semantic gap" on the shop floor. A deep learning model might correctly identify a 2-millimeter scratch on a machined aluminum housing with 99% confidence. But the model does not know *what* the housing is, *where* the scratch is located relative to critical mating surfaces, or *whether* a 2mm scratch violates the specific engineering tolerances for that particular SKU. Consequently, the system simply flags the part as a defect, halts the line, and waits for a human quality engineer to walk over, inspect the part, consult the CAD drawings, and make a disposition decision. This human-in-the-loop bottleneck negates the speed advantages of automated inspection and drives up the Cost of Poor Quality (COPQ) through excessive false positives and unnecessary scrap.

The **Vision Defect Detector** pattern fundamentally resolves this semantic gap by elevating the camera from a passive sensor to an active, reasoning agent. By leveraging multimodal Large Language Models (LLMs)—such as Gemini 3.5 Pro or Gemini 3.5 Flash-Lite—the agentic vision system does not merely detect an anomaly; it interprets the anomaly within the broader context of the manufacturing environment. It understands the physical geometry of the part, retrieves the relevant engineering specifications via Retrieval-Augmented Generation (RAG), reasons about the severity of the defect, and autonomously invokes tools to execute a disposition strategy.

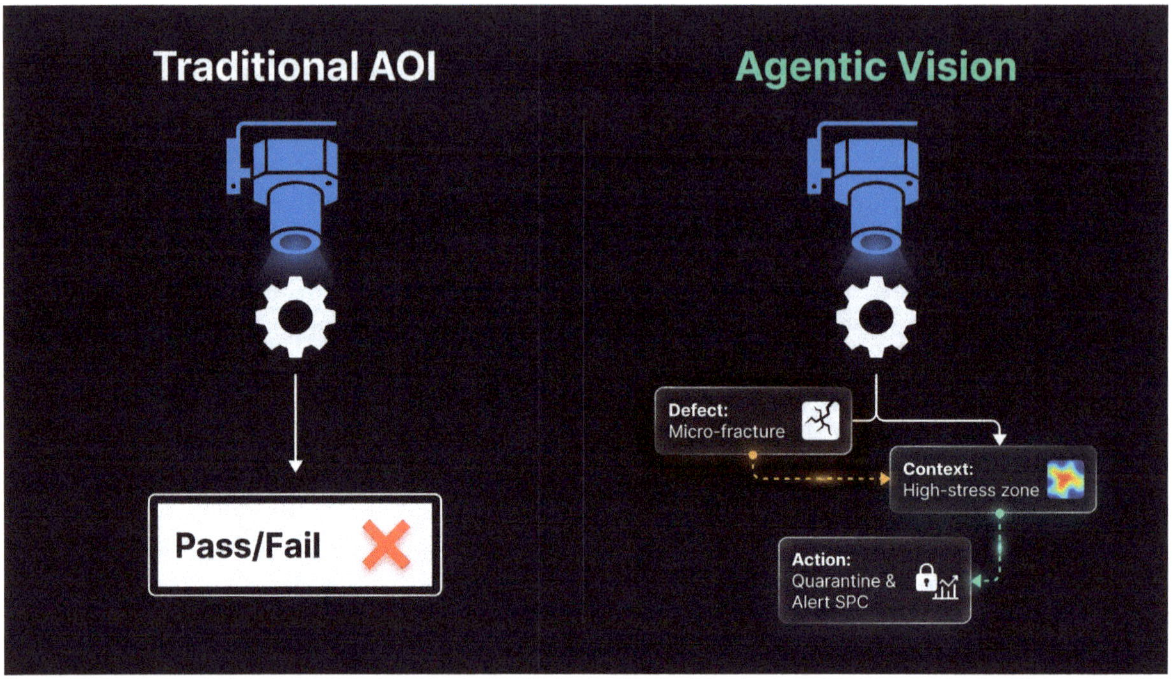

Figure 1.34: The paradigm shift from binary Pass/Fail detection in traditional AOI to semantic reasoning and autonomous action in Agentic Vision.

1.61.1 Architecting the Cascade Inference Pipeline

The primary challenge in deploying agentic vision on a manufacturing line is latency. High-speed discrete manufacturing lines (e.g., bottling, PCB surface mount technology, or battery cell winding) often produce parts at a rate of several units per second. The latency budget for an inline inspection decision is typically between 50 and 200 milliseconds. Multimodal LLMs, even highly optimized edge variants, cannot currently process high-resolution images and generate complex reasoning tokens within a 50ms window.

To reconcile the need for deep reasoning with the physics of high-speed manufacturing, architects must implement a **Cascade Inference Pipeline**. This architecture decouples the *detection* phase from the *reasoning* phase.

1. **The Trigger Layer (Edge):** A highly optimized, lightweight object detection model (such as YOLOv10 or a distilled MobileNet) runs directly on the edge Industrial PC (IPC) connected to the GigE Vision cameras. This model operates at 10-20ms latency. Its sole job is to act as an anomaly trigger. It does not classify the defect or make a disposition decision; it simply identifies regions of interest (ROIs) that deviate from the norm. If the part is perfect, the trigger model passes it, and the agent is never invoked.
2. **The Reasoning Layer (Edge/Local Server):** When the trigger model detects an anomaly, it crops the ROI and sends it, along with the full-frame context image, to the Vision Defect

Agent. This agent is powered by a multimodal LLM (e.g., Gemini Flash deployed on a local edge server or via a low-latency cloud VPC). Because the trigger model filters out 95% of the nominal parts, the Vision Agent only processes the 5% of parts that require complex reasoning. This allows the system to absorb the 500ms-1000ms latency of the LLM without slowing down the overall line throughput, utilizing buffer zones or parallel inspection stations to manage the asynchronous reasoning workload.

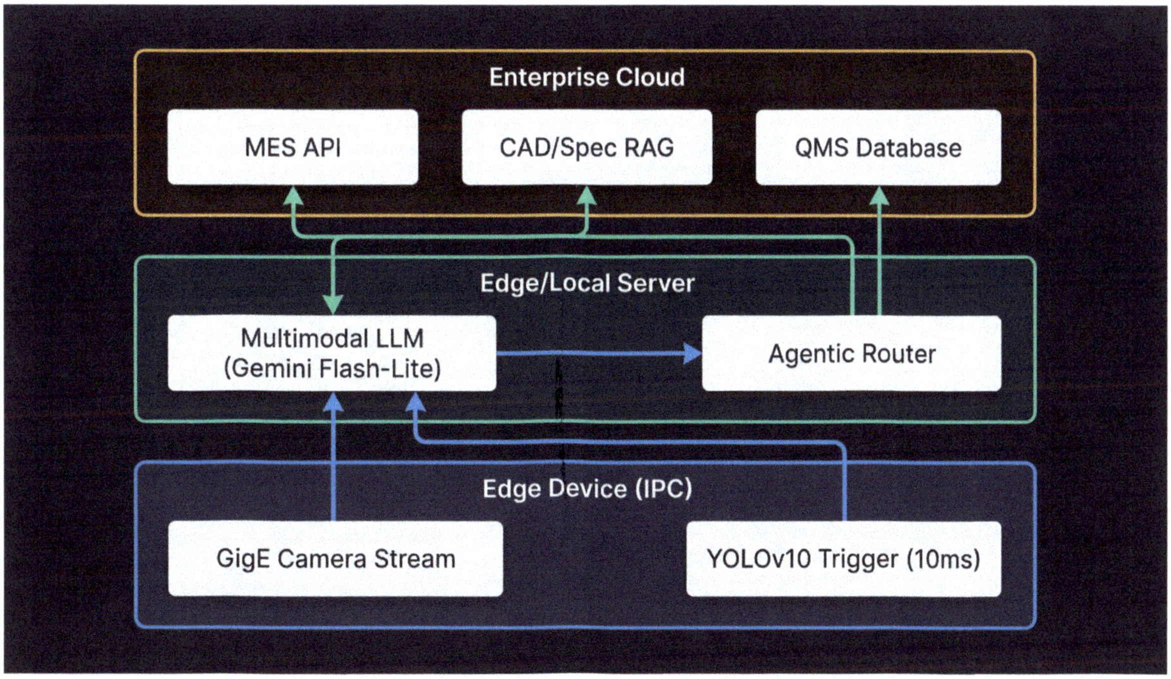

Figure 1.35: Cascade Architecture: A lightweight edge model triggers the heavier multimodal reasoning agent only when anomalies are detected, preserving line speed.

1.61.2 Contextual Grounding and Tool Invocation

Once the Vision Defect Agent receives an image of an anomaly, it must determine the severity and the appropriate action. To do this accurately, the agent cannot rely solely on its pre-trained weights; it must be grounded in the specific reality of the factory floor. This is achieved through tool invocation and Context Engineering.

When the agent is invoked, it is provided with a system prompt that defines its persona (e.g., "You are an expert metallurgical quality engineer") and a set of available tools. The typical execution loop follows these steps:

1. **Identify the Part:** The agent reads the barcode or QR code in the image (or receives the Part ID from the PLC metadata) to identify the specific SKU.
2. **Retrieve Specifications (Tool Call):** The agent calls a `get_engineering_specs(sku_id)` tool. This tool queries a vector database containing the CAD models, engineering drawings, and acceptable tolerance limits for that specific part.
3. **Multimodal Analysis:** The agent analyzes the image of the defect against the retrieved specifications. For example, it might reason: *"The image shows a 1.5mm porosity defect on*

the flange. The retrieved specification for SKU-882 states that porosity up to 2.0mm is acceptable on the flange, provided it is not within 5mm of the bolt hole. The image shows the defect is 12mm away from the bolt hole."

4. **Execute Disposition (Tool Call):** Based on its reasoning, the agent makes a decision. It calls the `update_mes_disposition(part_id, status, reason)` tool. In this case, it logs the defect for tracking but marks the part as "Pass/Acceptable Deviation," preventing unnecessary scrap.

Vision Defect Detector Evolution

	Traditional CV	Deep Learning CV	Agentic Vision
Detection Method	Pixel Rules	Pattern Matching	Semantic Reasoning
Context Awareness	None	Low (Bounding Boxes)	High (RAG + Specs)
Action Space	Pass/Fail Signal	API Webhook	Tool Invocation (MES/QMS)
Adaptability	Rigid/Manual	Requires Retraining	Zero-Shot / Few-Shot

Figure 1.36: Capability Matrix illustrating the evolution of computer vision in manufacturing quality assurance.

1.61.3 Implementation: Building the Vision Agent

To implement this pattern reliably, we must force the multimodal LLM to output structured data. If the agent responds with a conversational paragraph, the downstream MES cannot parse the disposition command. We utilize the structured output capabilities of modern models (like Gemini's `response_schema` or OpenAI's structured outputs) to guarantee a JSON payload that conforms to a strict Pydantic model.

Below is a production-grade implementation using Python, Pydantic, and the Google GenAI SDK. This code defines the schema, sets up the agent with a strict system instruction, and processes an incoming defect image.

Listing 1.37: Implementing a structured Vision Defect Agent using Pydantic and Multimodal LLMs.

```
import os
from google import genai
from google.genai import types
```

```python
from pydantic import BaseModel, Field
from typing import List, Optional

# 1. Define the strict output schema for the Agent
class DefectAnalysis(BaseModel):
    defect_detected: bool = Field(description="True if a defect is present, False otherwise.")
    defect_type: Optional[str] = Field(description="Classification of the defect (e.g., Scratch, Porosity, Solder Bridge).")
    severity_score: int = Field(description="Severity from 1 (cosmetic) to 5 (critical failure).", ge=1, le=5)
    semantic_description: str = Field(description="Detailed reasoning explaining the defect and its location relative to critical features.")
    recommended_action: str = Field(description="Must be one of: PASS, REWORK, SCRAP, QUARANTINE.")
    requires_human_review: bool = Field(description="True if severity is >= 4 or confidence is low.")

# 2. Initialize the client and model
client = genai.Client(api_key=os.environ.get("GEMINI_API_KEY"))
MODEL_ID = "gemini-2.5-flash" # Optimized for speed and multimodal reasoning

def analyze_defect(image_path: str, part_sku: str, tolerance_spec: str) -> DefectAnalysis:
    """
    Analyzes an image of a part and returns a structured disposition decision.
    """
    # Upload the image to the Gemini API
    image_file = client.files.upload(file=image_path)

    # Construct the contextual prompt
    system_instruction = """
    You are an expert metallurgical quality assurance agent.
    Your job is to analyze images of machined parts, identify defects, and make disposition decisions based strictly on the provided engineering tolerances.
    You must output valid JSON conforming to the requested schema.
    """

    user_prompt = f"""
    Analyze the attached image for part SKU: {part_sku}.

    ENGINEERING TOLERANCE SPECIFICATION:
    {tolerance_spec}

    Determine if the visible anomaly violates the specification. Provide a semantic description of the defect's location and recommend an action.
    """

    # Call the model with structured output enforcement
    response = client.models.generate_content(
        model=MODEL_ID,
        contents=[image_file, user_prompt],
        config=types.GenerateContentConfig(
```

```python
            system_instruction=system_instruction,
            response_mime_type="application/json",
            response_schema=DefectAnalysis,
            temperature=0.1 # Low temperature for deterministic QA decisions
        ),
    )

    # Parse the JSON response into our Pydantic model
    return DefectAnalysis.model_validate_json(response.text)

# Example Execution
if __name__ == "__main__":
    # In a real system, this spec is retrieved via a RAG tool call
    spec = "SKU-882 Flange: Surface scratches < 2mm length are acceptable. Scratches crossing the O-ring groove are CRITICAL and must be SCRAPPED."

    analysis = analyze_defect("trigger_crop_001.jpg", "SKU-882", spec)

    print(f"Action: {analysis.recommended_action}")
    print(f"Reasoning: {analysis.semantic_description}")

    # Downstream routing logic based on the agent's structured output
    if analysis.recommended_action == "SCRAP":
        # trigger_mes_scrap_routine(part_id="SKU-882")
        pass
```

1.61.4 Handling Edge Cases and Hallucinations

While multimodal LLMs are powerful, they are susceptible to hallucinations—particularly in industrial settings where lighting variations, glare on metallic surfaces, or oil residue can mimic the appearance of defects. A robust Vision Defect Detector must implement defensive engineering patterns to mitigate these risks.

Multi-Angle Verification: A single 2D image can be deceptive. A scratch might look like a deep gouge depending on the angle of the key light. Advanced agentic systems utilize a "Multi-Angle Verification" pattern. If the agent detects a critical defect with a severity score of 4 or 5, it does not immediately scrap the part. Instead, it invokes a tool to control the robotic inspection arm or the multi-camera array: `request_new_image(angle=45, lighting="darkfield")`. The agent then analyzes the new image to confirm the hypothesis before executing the final disposition.

Confidence Thresholds and Human Handoff: The `requires_human_review` boolean in our Pydantic schema is a critical safety valve. The agent is programmed to set this flag to `True` if the defect characteristics fall into a gray area not explicitly covered by the retrieved engineering specifications, or if the visual evidence is ambiguous. When this flag is triggered, the agent routes the image, the retrieved specs, and its preliminary reasoning to a human quality engineer's dashboard. This ensures that the AI handles the 90% of clear-cut cases, while humans handle the complex edge cases, drastically improving overall throughput without compromising quality standards.

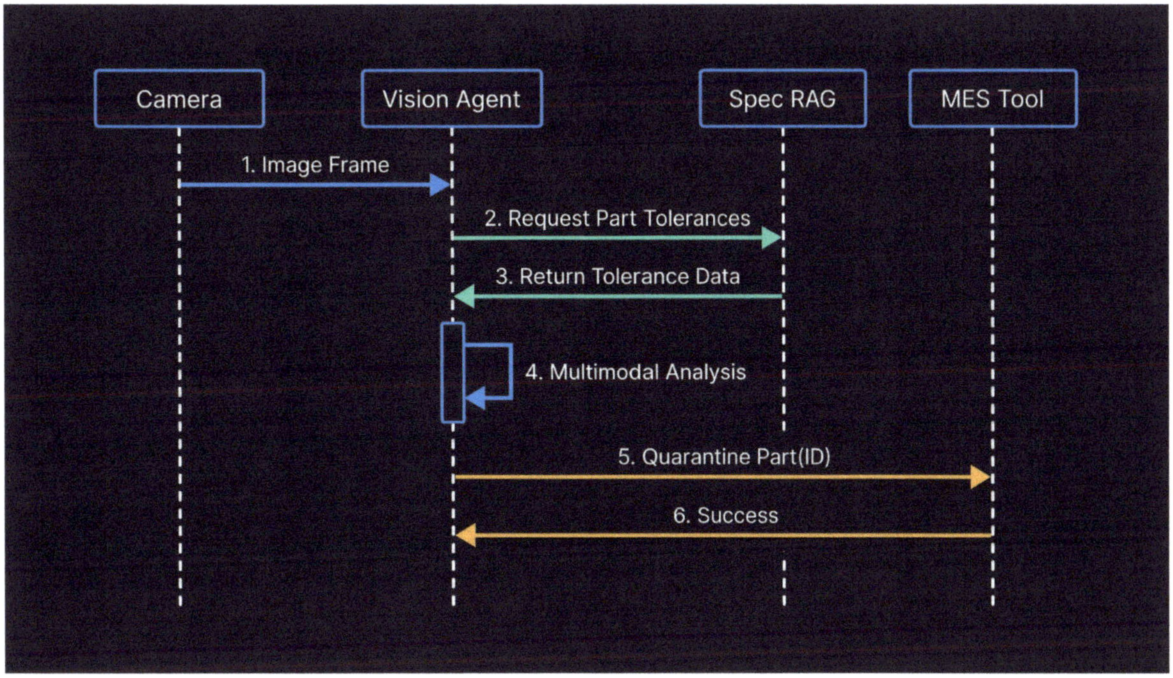

Figure 1.37: Sequence diagram detailing the autonomous tool-calling workflow of the Vision Defect Agent.

1.61.5 Integration with the Broader Agentic Mesh

The Vision Defect Detector does not operate in a vacuum; it is the sensory input for the broader multi-agent Quality Management System. When the vision agent confirms a defect, it publishes an event to the central message broker (e.g., Kafka or Google Cloud Pub/Sub).

This event triggers downstream agents. For instance, if the vision agent detects a recurring dimensional anomaly (e.g., three consecutive parts with undersized flanges), it alerts the **SPC Anomaly Agent**. The SPC agent will then autonomously query the machine telemetry (spindle speed, vibration, temperature) to determine if tool wear is the root cause. Simultaneously, the **Traceability Copilot** is triggered to map the genealogy of the defective parts, identifying which supplier provided the raw aluminum billet, thereby closing the loop from edge detection to enterprise-wide root cause analysis.

Key Takeaways

- **Semantic reasoning over binary classification:** Agentic vision moves beyond simple bounding boxes, allowing the system to understand the context, location, and severity of a defect relative to engineering specifications.
- **Cascade architectures are mandatory for inline speed:** Use lightweight, low-latency models (YOLO) at the edge to trigger the heavier, multimodal LLMs only when anomalies are detected, preserving line throughput.
- **Enforce structured outputs:** Always use schema enforcement (like Pydantic) to ensure the vision agent returns deterministic JSON that can be safely parsed by downstream MES and QMS APIs.

- **Design for uncertainty:** Implement multi-angle verification tools and strict Human-in-the-Loop (HITL) handoff protocols for ambiguous defects to prevent costly false positives and AI hallucinations.
- **Agents are sensors for the mesh:** The vision agent's output should serve as the trigger for downstream reasoning agents, such as SPC analyzers and Traceability copilots, to achieve closed-loop quality control.

1.62 SPC Anomaly Agent

Learning Objectives

- Understand the architectural shift from passive Statistical Process Control (SPC) to active, agentic process orchestration.
- Design a Cascade Inference Pipeline that separates high-frequency telemetry triggering from low-frequency LLM reasoning.
- Implement a LangGraph-based state machine for autonomous root cause analysis and closed-loop parameter adjustment.
- Establish safety guardrails and bounded action spaces for agents interacting directly with industrial control systems.

As established in the preceding section, the Vision Defect Agent serves as the sensory cortex of the modern manufacturing line. However, detecting a defect is only the first half of the quality assurance equation. When the vision agent confirms a recurring anomaly—such as three consecutive parts exhibiting an undersized flange—it publishes an event to the central message broker. This event acts as the catalyst for the **SPC Anomaly Agent**, a reasoning engine designed to transition quality control from reactive sorting to proactive, autonomous process correction.

Statistical Process Control (SPC) has been the bedrock of manufacturing quality since Walter A. Shewhart introduced control charts at Bell Labs in the 1920s. Traditional SPC relies on plotting process variables (like temperature, pressure, or dimensions) against statistically derived Upper and Lower Control Limits (UCL/LCL). When a data point falls outside these limits, or violates established heuristic patterns (such as the Nelson Rules), the system triggers an alarm.

The fundamental limitation of traditional SPC is that it is entirely passive. It tells an operator *that* a process is out of control, but it cannot explain *why*, nor can it fix it. In a modern Industry 4.0 facility generating terabytes of telemetry data per day, this passive alerting paradigm leads to severe alarm fatigue. Operators are inundated with flashing red lights on SCADA screens, forcing them to manually cross-reference historian data, maintenance logs, and tooling schedules to deduce the root cause.

The SPC Anomaly Agent fundamentally alters this dynamic. By wrapping a Large Language Model (LLM) in an agentic framework equipped with specific industrial tools, we create a system that not only detects the statistical deviation but autonomously investigates the root cause, synthesizes a hypothesis, and—crucially—executes corrective actions back to the machine controller.

1.62.1 Architecting the Agentic SPC Pipeline

Deploying an LLM to monitor high-frequency industrial telemetry presents an immediate architectural challenge: impedance mismatch. A CNC machine's spindle vibration sensor might sample at 10,000 Hz. An LLM cannot, and should not, process data at this frequency. Doing so would be computationally ruinous and entirely unnecessary, as the vast majority of the data represents nominal operation.

To resolve this, we employ a **Cascade Inference Pipeline**, similar to the pattern used in the Vision Defect Agent, but adapted for time-series data. This architecture strictly decouples the high-frequency *detection* phase from the low-frequency *reasoning* phase.

The architecture consists of three distinct layers:

1. **The Data Ingestion Layer:** Machine PLCs and IoT sensors stream telemetry via industrial protocols like OPC UA or MQTT. This data is ingested into a Time-Series Database (Historian) such as InfluxDB or OSIsoft PI. 2. **The Trigger Layer:** Lightweight, traditional machine learning models (e.g., Isolation Forests, Autoencoders, or even simple rolling-window Nelson Rule algorithms) run continuously against the incoming data streams. These models are computationally cheap and operate in near real-time. When they detect a statistically significant anomaly, they generate an event payload containing the timestamp, the anomalous variable, and a window of surrounding context data. 3. **The Reasoning Layer (The Agent):** The event payload is published to a message broker (like Apache Kafka), which wakes up the SPC Anomaly Agent. The agent, powered by a frontier model like Gemini 3.5 Pro, takes the anomaly event as its initial prompt and begins its investigation loop.

1.62.2 Contextual Reasoning and Data Fusion

When the SPC Anomaly Agent is invoked, it possesses only the symptom (e.g., "Spindle vibration exceeded UCL at 14:02:00"). To diagnose the disease, it must fuse multi-modal data from across the enterprise. This is where the agent's tool-calling capabilities become paramount.

The agent is equipped with a specific set of tools designed for industrial forensics:

- `query_historian(tag_id, start_time, end_time)`: Retrieves historical time-series data for specific machine parameters.
- `get_maintenance_logs(machine_id)`: Uses Retrieval-Augmented Generation (RAG) to search the Enterprise Asset Management (EAM) system for recent work orders, tool changes, or operator notes.
- `get_vision_events(machine_id, time_window)`: Queries the Vision Defect Agent's database to see if the telemetry anomaly correlates with physical defects on the parts being produced.

Consider a scenario where the trigger model flags an anomaly in the torque of a robotic screwdriving cell. The agent wakes up and begins its reasoning loop. First, it calls `query_historian` to look at the torque curve for the last 50 cycles. It notices a gradual upward trend rather than a sudden spike. Next, it calls `get_vision_events` and discovers that the Vision Agent has recently flagged two parts for "improper seating." Finally, it calls `get_maintenance_logs` and finds that a new batch of screws from a secondary supplier was loaded onto the line 45 minutes ago.

The LLM synthesizes this information: *"The gradual increase in torque, correlated with vision defects showing improper seating, immediately following a change in fastener batch, strongly suggests that the new screws have a slightly larger thread pitch or different coating friction. The root cause is likely supplier material variance, not a mechanical failure of the screwdriving spindle."*

This level of semantic reasoning—connecting a time-series trend to a text-based maintenance log and a computer vision event—is entirely impossible for traditional SPC software.

1.62.3 Implementation: Building the SPC Agent with LangGraph

To orchestrate this complex, multi-step reasoning process, we utilize LangGraph. LangGraph allows us to define the agent's workflow as a state machine, ensuring that it systematically gathers data, forms a hypothesis, and decides on an action without getting stuck in infinite loops.

Below is a production-grade implementation of the SPC Anomaly Agent's core reasoning graph. We define a state object to hold the context as it moves through the nodes, and we utilize conditional edges to route the agent based on its findings.

Listing 1.38: Implementing the SPC Anomaly Agent using LangGraph for stateful reasoning.

```python
from typing import TypedDict, Annotated, Sequence
from langchain_core.messages import BaseMessage, HumanMessage, AIMessage
from langgraph.graph import StateGraph, END
from langchain_google_genai import ChatGoogleGenerativeAI
import json

# 1. Define the State
class AgentState(TypedDict):
    messages: Annotated[Sequence[BaseMessage], "The conversation history"]
    anomaly_event: dict
    historian_data: str
    maintenance_context: str
    hypothesis: str
    recommended_action: str

# 2. Initialize the LLM (Gemini 3.5 Pro for complex reasoning)
llm = ChatGoogleGenerativeAI(model="gemini-1.5-pro", temperature=0.1)

# 3. Define the Nodes
def ingest_anomaly(state: AgentState):
    """Parses the incoming Kafka event and formats the initial prompt."""
    event = state["anomaly_event"]
    prompt = f"ANOMALY DETECTED: Machine {event['machine_id']}, Tag {event['tag']}, Value {event['value']} exceeded UCL at {event['timestamp']}. Begin investigation."
    return {"messages": [HumanMessage(content=prompt)]}

def gather_context(state: AgentState):
    """Simulates tool calls to Historian and EAM systems."""
    # In production, these would be actual API calls via LangChain Tools
    machine_id = state["anomaly_event"]["machine_id"]

```

```
    # Mock data retrieval
    historian_data = f"Trend for {machine_id}: Gradual 15% increase in vibration over last 2 hours."
    maintenance_context = f"Log for {machine_id}: Cutting tool replaced 4 hours ago. Operator noted 'tool seating felt tight'."

    context_msg = f"Historian Data: {historian_data}\nMaintenance Logs: {maintenance_context}"
    return {
        "historian_data": historian_data,
        "maintenance_context": maintenance_context,
        "messages": state["messages"] + [AIMessage(content=f"Gathered context: {context_msg}")]
    }

def analyze_root_cause(state: AgentState):
    """Uses the LLM to synthesize the data and form a hypothesis."""
    prompt = f"""
    Based on the following data, determine the root cause of the anomaly.
    Anomaly: {state['anomaly_event']}
    Historian: {state['historian_data']}
    Maintenance: {state['maintenance_context']}

    Provide a concise hypothesis and a recommended action (e.g., ADJUST_PARAMETER, SCHEDULE_MAINTENANCE, IGNORE).
    Format as JSON: {{"hypothesis": "...", "action": "..."}}
    """
    response = llm.invoke(state["messages"] + [HumanMessage(content=prompt)])

    # Parse JSON response
    try:
        result = json.loads(response.content.strip('json\n').strip(''))
        return {
            "hypothesis": result["hypothesis"],
            "recommended_action": result["action"],
            "messages": state["messages"] + [AIMessage(content=response.content)]
        }
    except:
        return {"hypothesis": "Failed to parse", "recommended_action": "HUMAN_REVIEW"}

# 4. Define the Graph
workflow = StateGraph(AgentState)

workflow.add_node("ingest", ingest_anomaly)
workflow.add_node("gather", gather_context)
workflow.add_node("analyze", analyze_root_cause)

workflow.set_entry_point("ingest")
workflow.add_edge("ingest", "gather")
workflow.add_edge("gather", "analyze")
workflow.add_edge("analyze", END)
```

```
# Compile the graph
spc_agent = workflow.compile()

# Example Execution
if __name__ == "__main__":
    initial_state = {
        "anomaly_event": {
            "machine_id": "CNC-04",
            "tag": "spindle_vibration",
            "value": 4.2,
            "timestamp": "2026-06-22T09:38:00Z"
        },
        "messages": [],
        "historian_data": "",
        "maintenance_context": "",
        "hypothesis": "",
        "recommended_action": ""
    }

    final_state = spc_agent.invoke(initial_state)
    print(f"Hypothesis: {final_state['hypothesis']}")
    print(f"Action: {final_state['recommended_action']}")
```

1.62.4 Autonomous Action and Closed-Loop Control

The ultimate goal of the SPC Anomaly Agent is not just to generate a report, but to close the loop by taking corrective action. If the agent determines that a dimensional drift is caused by tool wear, it should autonomously adjust the CNC machine's tool wear offset parameters via the PLC.

However, allowing an AI agent to write parameters directly to industrial control systems introduces significant safety and operational risks. A hallucinated parameter change could crash a spindle, destroy a part, or endanger human operators. Therefore, the architecture must implement strict **Bounded Action Spaces** and **Policy-as-Code Guardrails**.

Table~1.4 contrasts the passive alerting of traditional SPC with the autonomous, closed-loop control of the agentic approach.

Table 1.4: From passive alerting to autonomous closed-loop control: Traditional SPC versus the Agentic SPC Anomaly Agent.

Dimension	Traditional SPC	Agentic SPC Anomaly Agent
Detects deviation	Yes — flags points outside UCL/LCL or Nelson Rule violations	Yes — lightweight models trigger on statistical deviation
Explains root cause	No — reports *that* a process is out of control, not *why*	Yes — autonomously investigates and synthesizes a hypothesis
Corrective action	No — cannot fix the process	Yes — executes bounded corrections back to the machine controller
Operator burden	Manual cross-referencing of logs; severe alarm fatigue	Autonomous investigation; human only for escalated approvals
Variable scope	Typically univariate (one variable at a time)	Multivariate via PCA/autoencoders across many coupled variables

When the agent decides on an action (e.g., `adjust_plc_parameter(tag="tool_offset_x", value=0.05)`), the command does not go directly to the machine. Instead, it passes through an Agent Gateway or a Safety PLC layer. This layer enforces deterministic, hard-coded rules:

- **Range Limits:** The requested value must fall within a pre-approved safe operating envelope (e.g., offset adjustments cannot exceed ± 0.1mm per cycle).
- **Rate Limits:** The agent cannot make more than one adjustment per 15 minutes, preventing oscillatory over-correction.
- **Human-in-the-Loop (HITL) Escalation:** If the agent's confidence score is low, or if the requested action involves a critical parameter (like spindle speed), the gateway intercepts the command and routes it to a human supervisor's dashboard for one-click approval.

1.62.5 Advanced Pattern Recognition: Multivariate Agentic Control

Traditional SPC typically monitors one variable at a time (univariate analysis). However, modern manufacturing processes are highly coupled. A slight increase in ambient humidity might not trigger a univariate alarm, and a slight decrease in oven temperature might not trigger an alarm. But together, they might cause a catastrophic failure in a curing process.

Advanced SPC Anomaly Agents leverage the LLM's ability to interpret high-dimensional data. The trigger layer can utilize Principal Component Analysis (PCA) or Autoencoders to monitor the latent space of dozens of variables simultaneously. When the multivariate model detects a drift in the latent space, it passes the state vector to the agent.

The agent, equipped with a prompt that explains the physical relationships between the variables, can reason about the multivariate drift. It might conclude: *"The autoencoder flagged an anomaly. Reviewing the state vector, I see Variable A (Humidity) is at the 85th percentile and Variable B (Temperature) is at the 15th percentile. While neither violates their individual UCL/LCL, this specific combination is known to cause incomplete curing based on historical RAG data. Action: Increase oven temperature by 2 degrees to compensate for the high humidity."*

This represents the pinnacle of Industry 4.0 quality assurance: a system that not only sees the defect but understands the complex, multi-variable physics of the process that created it, and takes autonomous action to ensure the next part is perfect.

Key Takeaways

- **Move beyond passive alerting:** Agentic SPC transforms quality control from a reactive monitoring system that causes alarm fatigue into a proactive, autonomous orchestration engine.
- **Implement Cascade Architectures:** Never feed raw, high-frequency telemetry directly into an LLM. Use lightweight statistical models (Isolation Forests, Autoencoders) at the edge to trigger the reasoning agent only when necessary.
- **Fuse multi-modal context:** The true power of the agent lies in its ability to correlate time-series telemetry with text-based maintenance logs and visual defect events to formulate accurate root cause hypotheses.
- **Enforce Bounded Action Spaces:** When allowing agents to write back to PLCs or SCADA systems, implement strict, deterministic guardrails (range limits, rate limits) and Human-in-the-Loop protocols for critical parameters to ensure operational safety.
- **Leverage LangGraph for stateful reasoning:** Use graph-based orchestration frameworks to ensure the agent follows a systematic, repeatable process of data gathering, hypothesis generation, and action execution.

Learning Objectives

- **Understand the architectural shift** from relational, siloed traceability systems to GraphRAG-powered, multi-modal genealogy networks.
- **Design a multi-agent copilot** capable of autonomously traversing complex Bills of Materials (BOM) and routing queries across ERP, MES, and QMS boundaries.
- **Implement automated "blast radius" calculations** to rapidly identify, isolate, and contain defective batches across the supply chain.
- **Deploy tool-calling agents** that synthesize structured database records (SQL/-Cypher) with unstructured supplier documentation (PDF Certificates of Analysis).

While the SPC Anomaly Agent discussed in the previous chapter successfully corrects manufacturing processes to prevent *future* defects, a critical, time-sensitive question immediately follows every quality excursion: what happens to the parts produced *before* the intervention? If a CNC spindle vibration anomaly went undetected for two hours, or if a supplier notifies you that a batch of raw resin was contaminated, the clock starts ticking. You must rapidly identify every affected component, sub-assembly, and finished good, locate them within your global supply chain, and quarantine them before they reach the customer.

This is the domain of Traceability and Genealogy. In modern Industry 4.0 environments, traceability is no longer just a regulatory checkbox for ISO 9001 or FDA compliance; it is a foundational capability for risk mitigation, targeted recalls, and continuous improvement. However, despite decades of digitization, tracing the genealogy of a complex product remains one of the most labor-intensive, error-prone tasks in manufacturing.

This chapter explores the design and implementation of a **Traceability & Genealogy Copilot**—an agentic system that replaces the manual "swivel-chair" integration of quality engineers with autonomous graph traversal, multi-system tool invocation, and semantic reasoning.

1.63 The Traceability Crisis in Industry 4.0

To understand why an agentic approach is necessary, we must first examine the structural failures of traditional traceability workflows. Manufacturing data is inherently fragmented. The journey of a single product spans multiple disparate systems, each speaking a different data language and operating on different temporal scales.

Consider the genealogy of a modern electric vehicle (EV) battery pack. The raw lithium and cobalt arrive with unstructured PDF Certificates of Analysis (CoAs) from tier-2 suppliers. These materials are processed into cells, tracked by batch numbers in a local Manufacturing Execution System (MES). The cells are assembled into modules, and modules into packs, with the parent-child relationships (the Bill of Materials, or BOM) managed in a Product Lifecycle Management (PLM) system. Finally, the finished pack is serialized and shipped, recorded in an Enterprise Resource Planning (ERP) system like SAP or Oracle.

When a thermal runaway event occurs in the field, the quality engineer must perform a **Backward Trace** (from the finished good back to the raw material) to find the root cause, followed immediately by a **Forward Trace** (from the identified bad raw material forward to all other finished goods) to execute a containment hold.

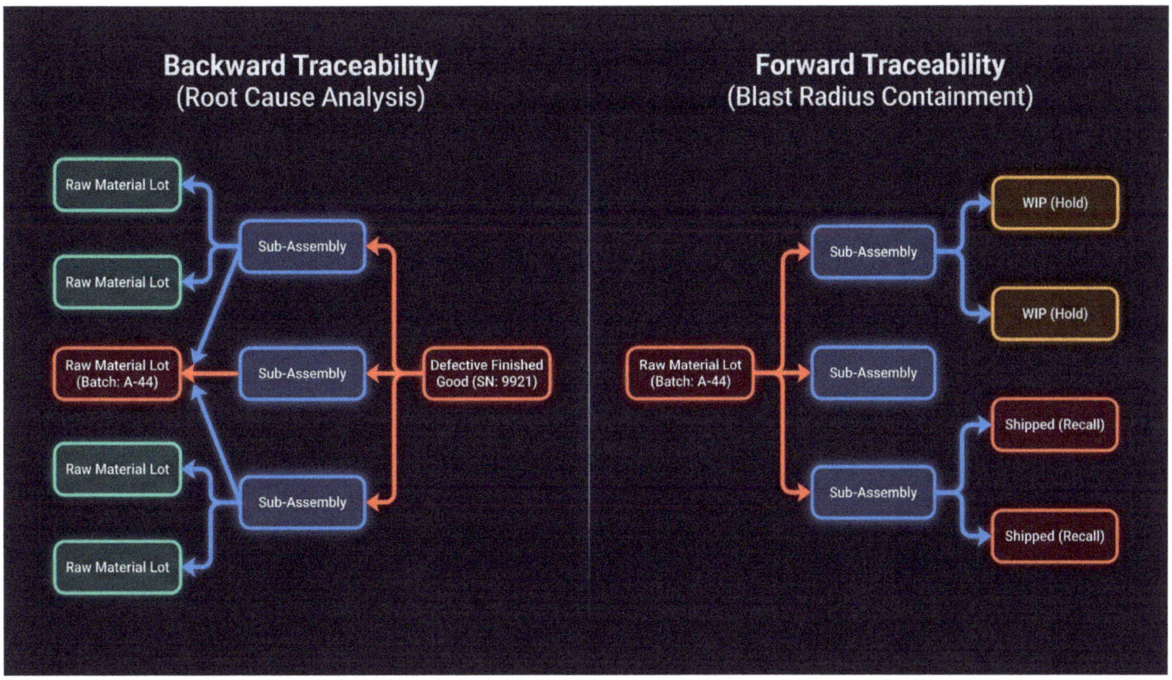

Figure 1.38: The Traceability Graph: Backward traceability identifies the root cause (e.g., a specific raw material batch), while forward traceability calculates the blast radius to contain all affected products.

In a traditional setup, this process is entirely manual. The engineer queries the ERP for the serial number, copies the sub-assembly IDs, pastes them into the MES to find the batch numbers, emails the supplier quality team for the CoAs, and manually builds a spreadsheet to map the blast radius. This process can take days or even weeks—time during which defective products might be installed in vehicles or shipped to consumers.

The core issue is that relational databases (SQL) are poorly suited for deep, recursive hierarchical queries (like traversing a multi-level BOM), and traditional software cannot easily bridge the gap between structured database rows and unstructured supplier documents.

1.64 Architecting the Genealogy Copilot

The Traceability & Genealogy Copilot solves this crisis by treating traceability not as a series of isolated database queries, but as an agentic reasoning task over a multi-modal knowledge graph. By equipping a Large Language Model (LLM) with specific tools to query graphs, relational databases, and document stores, we create an autonomous investigator capable of traversing the entire supply chain in seconds.

1.64.1 Core Architectural Components

The architecture of the Copilot relies on the **Planner-Executor-Evaluator** pattern, augmented with specialized retrieval mechanisms:

1. **The Orchestrator Agent:** The central reasoning engine (e.g., Gemini 3.5 Pro) that receives the initial anomaly report or serial number. It plans the investigation steps, deciding whether to trace forward or backward, and selects the appropriate tools.
2. **GraphRAG Tool (Text-to-Cypher):** To solve the BOM traversal problem, modern traceability systems mirror their relational data into Graph Databases (like Neo4j). The agent uses a tool that translates natural language intents ("Find all parent assemblies that consume Batch A-44") into graph query languages (Cypher), executing deep recursive searches instantly.
3. **ERP/MES SQL Tool (Text-to-SQL):** Once the graph identifies the affected serial numbers, the agent must determine their current physical location and status. It uses a Text-to-SQL tool to query the ERP/WMS (Warehouse Management System) to find if the parts are in WIP (Work in Progress), Finished Goods inventory, or already shipped.
4. **Document Retrieval Tool (Vector Search):** For supplier quality data, the agent queries a vector database containing embedded, OCR-processed Certificates of Analysis and supplier deviation reports.
5. **Action Execution Tool (API Integration):** The final, critical component. Once the blast radius is calculated, the agent uses an API tool to automatically place an electronic "Quality Hold" on the affected inventory in the ERP, preventing it from being shipped or consumed.

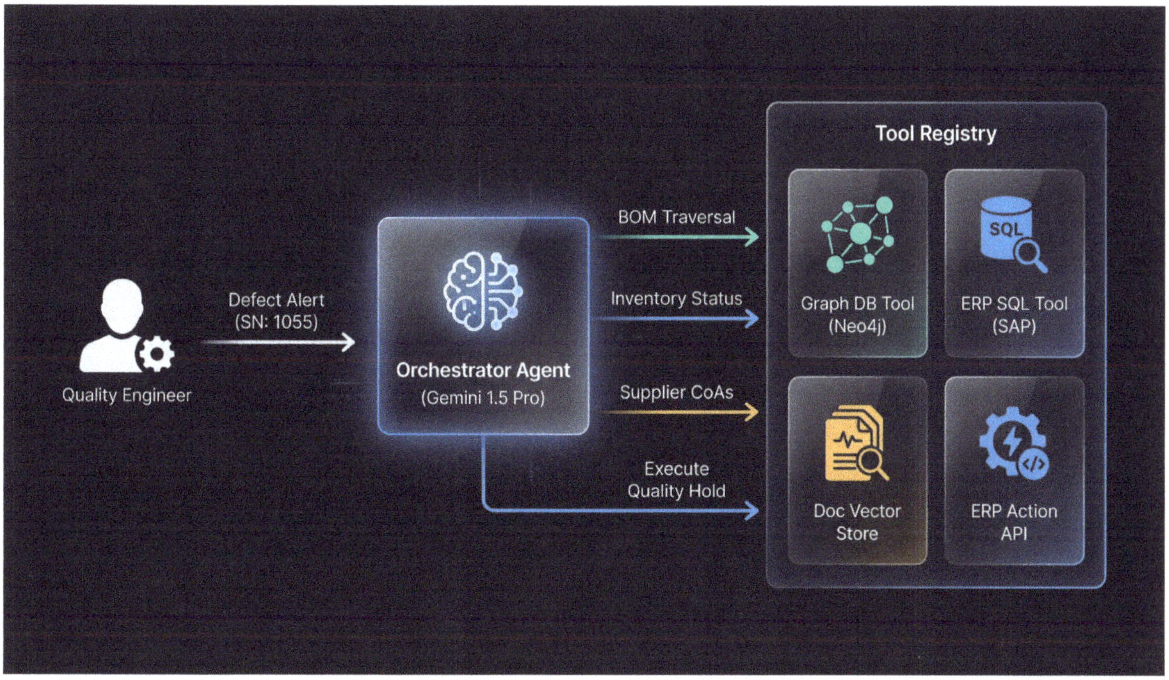

Figure 1.39: Architecture of the Traceability Copilot, demonstrating how the central agent orchestrates Graph, SQL, and Vector tools to synthesize a complete genealogy and execute containment actions.

1.65 Implementation: Building a Traceability Graph Agent

To demonstrate this pattern, we will implement a simplified Traceability Copilot using LangGraph. This agent will receive a defective serial number, query a mock Graph Database to find the root cause batch, trace forward to find all other affected serial numbers, and query an ERP to find their locations.

We utilize a stateful graph approach to ensure the agent follows a rigorous, auditable process rather than relying on a single, unpredictable zero-shot prompt.

Listing 1.39: LangGraph Implementation of a Traceability Copilot

```
from typing import TypedDict, Annotated, Sequence, List
from langchain_core.messages import BaseMessage, HumanMessage, AIMessage
from langgraph.graph import StateGraph, END
from langchain_google_genai import ChatGoogleGenerativeAI
import json

# 1. Define the Agent State
class TraceabilityState(TypedDict):
    messages: Annotated[Sequence[BaseMessage], "Conversation history"]
    trigger_sn: str
    root_cause_batch: str
    blast_radius_sns: List[str]
    inventory_status: dict
    containment_plan: str
```

```python

# 2. Initialize the LLM
llm = ChatGoogleGenerativeAI(model="gemini-1.5-pro", temperature=0.0)

# 3. Define Mock Tools (In production, these connect to Neo4j and SAP/ERP)
def tool_query_graph_backward(sn: str) -> str:
    """Simulates a Cypher query to find the raw material batch for a given SN."""
    # Mock graph traversal: (FinishedGood {id: sn})-[:CONSUMES*]->(RawMaterial)
    mock_db = {"SN-9921": "BATCH-A44", "SN-9922": "BATCH-A44", "SN-9923": "BATCH-B12"}
    return mock_db.get(sn, "UNKNOWN_BATCH")

def tool_query_graph_forward(batch: str) -> List[str]:
    """Simulates a Cypher query to find all SNs that consumed a specific batch."""
    # Mock graph traversal: (RawMaterial {id: batch})<-[:CONSUMES*]-(FinishedGood)
    mock_db = {"BATCH-A44": ["SN-9921", "SN-9922", "SN-9925", "SN-9926"]}
    return mock_db.get(batch, [])

def tool_query_erp_inventory(sns: List[str]) -> dict:
    """Simulates an SQL query to the ERP to find the location of serial numbers."""
    mock_erp = {
        "SN-9921": "SHIPPED - Customer X",
        "SN-9922": "WIP - Assembly Line 2",
        "SN-9925": "INVENTORY - Warehouse B",
        "SN-9926": "INVENTORY - Warehouse B"
    }
    return {sn: mock_erp.get(sn, "NOT_FOUND") for sn in sns}

# 4. Define Graph Nodes
def backward_trace_node(state: TraceabilityState):
    """Identifies the root cause batch from the defective serial number."""
    sn = state["trigger_sn"]
    batch = tool_query_graph_backward(sn)
    msg = f"Backward Trace Complete: SN {sn} was built using Raw Material {batch}."
    return {
        "root_cause_batch": batch,
        "messages": state["messages"] + [AIMessage(content=msg)]
    }

def forward_trace_node(state: TraceabilityState):
    """Calculates the blast radius by finding all SNs that used the bad batch."""
    batch = state["root_cause_batch"]
    affected_sns = tool_query_graph_forward(batch)
    msg = f"Forward Trace Complete: Batch {batch} was used in SNs: {affected_sns}."
    return {
        "blast_radius_sns": affected_sns,
```

```
        "messages": state["messages"] + [AIMessage(content=msg)]
    }

def locate_inventory_node(state: TraceabilityState):
    """Queries the ERP to find the physical location of all affected SNs."""
    sns = state["blast_radius_sns"]
    status = tool_query_erp_inventory(sns)
    msg = f"Inventory Status Retrieved: {json.dumps(status)}"
    return {
        "inventory_status": status,
        "messages": state["messages"] + [AIMessage(content=msg)]
    }

def draft_containment_node(state: TraceabilityState):
    """Uses the LLM to synthesize the findings and draft a containment plan.""
    "
    prompt = f"""
    You are a Quality Assurance Copilot. A defect was found in SN: {state['
    trigger_sn']}.
    Based on the traceability data, draft a containment plan.

    Root Cause Batch: {state['root_cause_batch']}
    Affected Serial Numbers (Blast Radius): {state['blast_radius_sns']}
    Current Locations: {json.dumps(state['inventory_status'], indent=2)}

    Provide specific instructions for WIP, Inventory, and Shipped goods.
    """
    response = llm.invoke(state["messages"] + [HumanMessage(content=prompt)])
    return {
        "containment_plan": response.content,
        "messages": state["messages"] + [AIMessage(content=response.content)]
    }

# 5. Build and Compile the Workflow
workflow = StateGraph(TraceabilityState)

workflow.add_node("backward_trace", backward_trace_node)
workflow.add_node("forward_trace", forward_trace_node)
workflow.add_node("locate_inventory", locate_inventory_node)
workflow.add_node("draft_containment", draft_containment_node)

workflow.set_entry_point("backward_trace")
workflow.add_edge("backward_trace", "forward_trace")
workflow.add_edge("forward_trace", "locate_inventory")
workflow.add_edge("locate_inventory", "draft_containment")
workflow.add_edge("draft_containment", END)

traceability_copilot = workflow.compile()

# 6. Execute the Copilot
if __name__ == "__main__":
    initial_state = {
        "trigger_sn": "SN-9921",
        "messages": [],
```

```
            "root_cause_batch": "",
            "blast_radius_sns": [],
            "inventory_status": {},
            "containment_plan": ""
        }

        result = traceability_copilot.invoke(initial_state)
        print("\n--- CONTAINMENT PLAN ---")
        print(result["containment_plan"])
```

In a production environment, the mock functions in the code above are replaced by robust LangChain Tools. The `tool_query_graph_backward` function would utilize a `GraphCypherQAChain` connected to a Neo4j instance, allowing the LLM to dynamically generate Cypher queries based on the specific schema of the manufacturer's BOM. Similarly, the ERP query would utilize an `SQLDatabaseToolkit` connected to a read-replica of the SAP HANA database.

1.66 Containment and Blast Radius Analysis

The most critical output of the Traceability Copilot is the **Blast Radius Analysis**. When a defect is confirmed, the goal is to minimize the financial and reputational damage by containing the issue as narrowly as possible. If traceability is poor, a manufacturer might be forced to recall an entire month's worth of production. With precise, agentic traceability, they can recall only the specific 40 units that consumed the contaminated resin.

Capability	Traditional Traceability	Agentic Copilot
Data Integration	Siloed (Manual Export)	Multi-modal (SQL, Graph, PDF)
BOM Traversal	Manual Relational Joins	Autonomous Cypher Generation
Blast Radius Calculation	Hours to Days	Instant / Automated
Containment Action	Manual Data Entry	API-driven ERP Holds

Figure 1.40: Comparison Matrix illustrating the operational advantages of an Agentic Traceability Copilot over traditional manual workflows.

The Copilot automates the blast radius calculation through a deterministic sequence:

1. **Identify the Contaminant:** Pinpoint the exact lot, batch, or serial number of the offending component.
2. **Traverse the Graph Forward:** Execute a recursive query to find all parent assemblies that incorporate the contaminant. This must account for complex manufacturing realities, such as batch splitting, rework loops, and alternative BOMs.
3. **Map to Inventory Status:** Cross-reference the resulting list of finished goods against the ERP's inventory ledger.
4. **Categorize Exposure:** The agent categorizes the risk into actionable buckets:
    - *WIP (Work in Progress):* Parts currently on the shop floor. The agent can trigger an alert to the MES to lock the specific workstations, preventing further value-add on doomed parts.
    - *Finished Goods (Warehouse):* Parts sitting in inventory. The agent executes an API call to change their ERP status from "Available" to "Quality Hold," preventing the warehouse management system from picking them for outbound shipments.
    - *In-Transit / Shipped:* Parts that have left the facility. The agent drafts a targeted recall notice, identifying the specific customer orders, shipping manifests, and destination addresses.

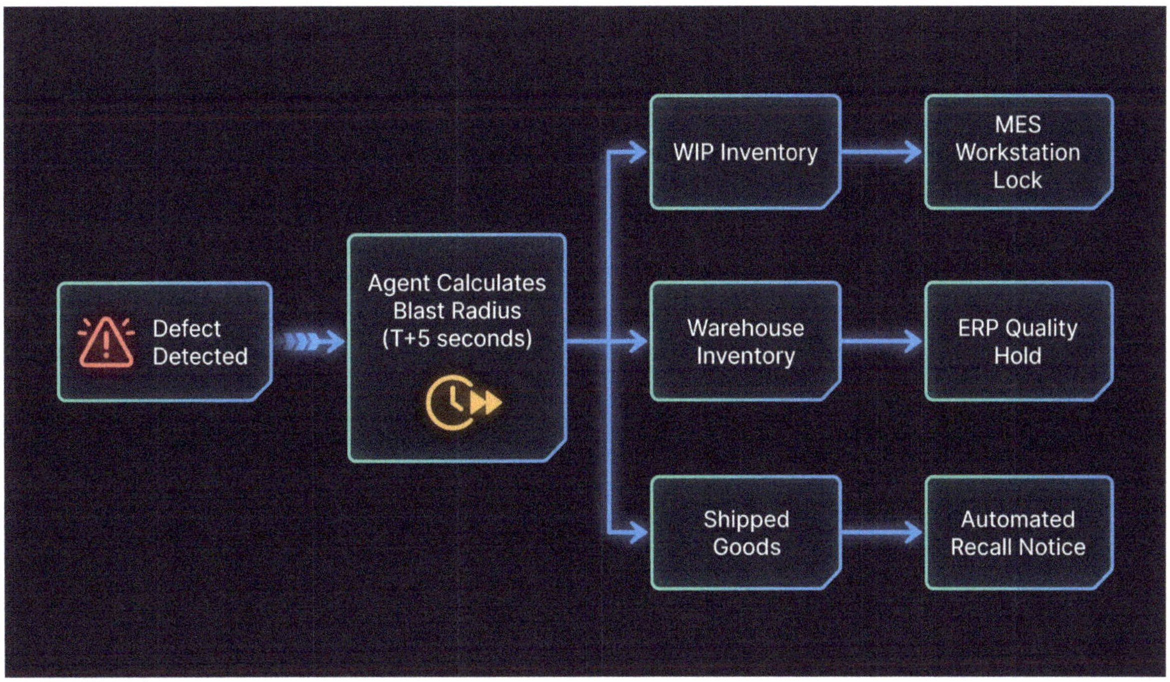

Figure 1.41: The automated containment workflow. By calculating the blast radius in seconds, the agent can execute system-level holds before defective products move further down the supply chain.

1.67 Supplier Quality and Multi-Tier Visibility

True traceability extends beyond the four walls of the factory. A significant percentage of quality excursions originate in the supply chain (Tier 1, Tier 2, and beyond). Integrating supplier data into the traceability graph has historically been a massive challenge because suppliers often provide data in unstructured formats—specifically, PDF Certificates of Analysis (CoAs) or email attachments.

This is where the multi-modal capabilities of frontier models like Gemini 3.5 Pro become transformative. The Traceability Copilot can be equipped with a **Document Ingestion Pipeline**. When a shipment of raw material arrives at the receiving dock, the accompanying PDF CoA is scanned.

Instead of a human manually typing the chemical composition or dimensional tolerances into the ERP, an ingestion agent processes the document. It extracts the supplier name, lot number, and key quality metrics, structuring them into JSON. This structured data is then written to the relational database, while the semantic content of the document is embedded into a Vector Store, and a node is created in the Graph Database linking the supplier lot to the internal receiving lot.

If a defect occurs weeks later, the Traceability Copilot can trace backward to the internal lot, traverse the graph to find the supplier lot, and use its Vector Search tool to retrieve the original CoA. The agent can then reason over the document: *"The backward trace points to Supplier Lot X. Reviewing the CoA for Lot X, I note that the tensile strength was recorded at 410 MPa, which is at the absolute lower limit of our acceptable tolerance. Combined with the high humidity recorded in our facility that day, this likely caused the structural failure."*

This level of automated, cross-silo, multi-modal reasoning represents a paradigm shift in Quality Assurance. It moves the organization from a reactive posture—spending days digging through filing cabinets and databases to answer "what happened?"—to a proactive, agent-driven posture where root causes are identified, and blast radii are contained, in a matter of seconds.

Key Takeaways

- **Graph Databases are essential:** Relational databases struggle with the recursive queries required for deep BOM traversal. Traceability copilots should leverage GraphRAG (Text-to-Cypher) to navigate complex parent-child relationships efficiently.
- **Multi-modal synthesis is the differentiator:** The agent's ability to combine structured ERP data (inventory locations) with unstructured supplier documents (PDF CoAs) provides a holistic view of the supply chain that traditional software cannot match.
- **Automate the containment, not just the search:** The highest ROI from a Traceability Copilot comes from integrating it with ERP/MES APIs to automatically execute Quality Holds, preventing defective WIP from advancing and defective inventory from shipping.
- **Stateful orchestration ensures reliability:** Use frameworks like LangGraph to enforce a deterministic sequence of operations (Backward Trace → Forward Trace → Locate → Contain), ensuring the agent does not skip critical investigative steps.

Learning Objectives

- Understand the structural complexities of Certificates of Analysis (COAs) and Production Part Approval Process (PPAP) documentation that cause traditional OCR systems to fail.
- Architect a multi-agent system capable of multimodal ingestion, semantic tolerance matching, and historical drift analysis.
- Implement a stateful, interruptible LangGraph workflow that integrates with PLM/ERP systems to validate supplier quality data.
- Design Human-in-the-Loop (HITL) governance models to ensure compliance with stringent manufacturing quality standards while maximizing automation ROI.

As established in the previous chapter on Traceability Copilots, the ability to rapidly traverse a Bill of Materials (BOM) and isolate defective components is a superpower for modern manufacturing. However, traceability is inherently reactive—it mitigates the blast radius *after* a defect has entered the facility. To achieve true operational excellence, manufacturers must shift left, preventing non-conforming materials from ever crossing the receiving dock.

This proactive defense relies entirely on Supplier Quality Assurance (SQA), a domain historically bottlenecked by the manual review of complex, unstructured documents: the Certificate of Analysis (COA) and the Production Part Approval Process (PPAP). In this chapter, we explore how agentic AI transforms this labor-intensive bottleneck into an automated, high-fidelity gateway, utilizing multimodal reasoning to validate supplier submissions against internal engineering specifications in seconds rather than days.

1.68 The Anatomy of Supplier Quality Bottlenecks

To appreciate the necessity of an agentic approach, one must first understand the sheer density and variability of the data involved in supplier quality validation. When a Tier 1 automotive supplier or an aerospace manufacturer receives a shipment of raw materials (e.g., a coil of aluminum alloy) or a batch of machined components, the physical goods are accompanied by a data payload.

For raw materials, this is typically a **Certificate of Analysis (COA)**. A COA is a document issued by the supplier detailing the specific chemical composition, physical properties, and dimensional measurements of that exact batch. It serves as a legal guarantee that the material meets the agreed-upon specifications. For complex manufactured components, the requirement is far more rigorous: the **Production Part Approval Process (PPAP)**. Originating in the automotive industry (via the AIAG standards) but now ubiquitous across discrete manufacturing, a PPAP is a comprehensive package of evidence proving that the supplier's manufacturing process is capable of consistently producing parts that meet the buyer's engineering design records.

A Level 3 PPAP submission contains 18 distinct elements, including the Part Submission Warrant (PSW), Design Failure Mode and Effects Analysis (DFMEA), Process Flow Diagrams, Control Plans, Measurement System Analysis (MSA), and dimensional results. A single PPAP package can easily exceed 100 pages of highly technical, heterogeneous data. It contains dense tables, engineering drawings with geometric dimensioning and tolerancing (GD&T) callouts, handwritten signatures, and statistical process control (SPC) charts.

1.68.1 The Failure of Traditional Automation

For the past decade, manufacturers have attempted to automate COA and PPAP ingestion using Optical Character Recognition (OCR) and Robotic Process Automation (RPA). These initiatives have almost universally failed to scale, achieving straight-through processing (STP) rates of less than 20%. The reasons are structural:

1. **Template Brittleness:** Traditional OCR relies on bounding boxes and coordinate mapping. If Supplier A moves the "Tensile Strength" field 2 centimeters to the left, the OCR pipeline breaks. Given that a large manufacturer might have 5,000 active suppliers, each using their own proprietary ERP system to generate COAs, maintaining OCR templates is an impossible task.

2. **Semantic Disconnects and Unit Conversions:** A supplier might report yield strength in *ksi* (kilopounds per square inch), while the manufacturer's internal Product Lifecycle Management (PLM) system stores the specification in *MPa* (Megapascals). Furthermore, the supplier might label the field "Yld Str," "Yield," or "Rp0.2". Traditional RPA cannot perform the semantic mapping required to understand that these are the same metric, nor can it dynamically execute the mathematical conversion required to validate the value against the specification.

3. **Multimodal Blind Spots:** PPAPs are not just text. They contain engineering drawings where critical dimensions are indicated by balloons (numbered circles pointing to features). The dimensional results table references these balloon numbers. Validating the PPAP requires looking at the drawing, finding balloon #4, reading the GD&T tolerance

(e.g., Ø10.5±0.1), and then checking the table to ensure the measured value is 10.45. OCR is entirely blind to this spatial, geometric reasoning.

1.68.2 The Cost of Poor Quality (COPQ)

Because automation has historically failed, SQA teams rely on manual review. A highly trained Quality Engineer must sit at a dual-monitor setup, opening the 100-page PDF on one screen and the SAP/Teamcenter interface on the other, manually "stare-and-compare" checking hundreds of data points. This process takes 4 to 8 hours per PPAP.

The consequences of this bottleneck are severe. First, it introduces massive supply chain latency; parts sit in quarantine for days waiting for approval. Second, it suffers from human fatigue. After reviewing 50 pages of dense tables, an engineer might miss that a Cpk (Process Capability Index) value is 1.32 instead of the required 1.33. If that part is approved and fails in the field, the resulting recall can cost tens of millions of dollars. This is the exact problem space where Agentic AI—specifically, multimodal models with massive context windows like Gemini 3.5 Pro—delivers transformative ROI.

1.69 Architecting the Digital COA/PPAP Reviewer

To solve the supplier quality bottleneck, we must move beyond simple text extraction and build a reasoning engine. The Digital COA/PPAP Reviewer is not a single prompt; it is a multi-agent architecture designed to mimic the cognitive workflow of a Lead Quality Engineer. It must ingest unstructured data, query internal systems of record for the ground-truth specifications, perform mathematical and semantic comparisons, and synthesize a final disposition.

1.69.1 The Multi-Agent Topology

The architecture utilizes a hierarchical, role-based multi-agent design. By separating concerns, we ensure that each agent has a narrow, well-defined system prompt and access only to the specific tools it needs, reducing hallucinations and improving deterministic execution.

The system comprises four primary agents:

1. The Ingestion & Triage Agent: This agent acts as the front door. When a PDF is uploaded to the supplier portal, this agent receives the file. Its primary job is classification and structured extraction. It uses a multimodal model (e.g., Gemini 3.5 Pro) to read the entire document natively. It identifies whether the document is a simple COA or a complex PPAP. It then extracts the core metadata: Supplier Name, Part Number, Batch/Lot Number, and Date. Crucially, it uses structured output (JSON schema enforcement) to extract the actual test results into a standardized key-value format, regardless of how the supplier formatted their tables.

2. The Specification Matching Agent: Once the data is structured, this agent takes over. It is equipped with API tools to query the manufacturer's internal systems. It takes the extracted Part Number and queries the PLM system (e.g., Siemens Teamcenter, PTC Windchill) to retrieve the active engineering specification for that exact revision of the part. It then performs *Semantic Tolerance Matching*. If the PLM says the hardness must be "HRC 40-45" and the extracted COA says "Hardness (Rockwell C): 42", the agent evaluates this as a PASS. If unit conversion is required, the agent utilizes a Python REPL tool to execute the math deterministically, rather than relying on the LLM's internal arithmetic capabilities.

3. The Historical Context Agent: Quality is not just about the current batch; it is about process stability over time. This agent utilizes GraphRAG and Vector Search to query the historical performance of this specific supplier and part. It looks for drift. For example, it might note: *"While the current batch's tensile strength of 410 MPa is within the acceptable tolerance of 400-450 MPa, this is the third consecutive batch where the value has decreased. The supplier's process is trending toward the lower control limit."* This insight is invaluable for predictive quality.

4. The Lead Quality Meta-Agent: This is the orchestrator. It receives the structured extraction, the specification validation results, and the historical context. Its job is to synthesize a final disposition report. It generates a confidence score for its own assessment. If all values are well within tolerance, the data is clear, and the confidence score is high (e.g., > 0.95), it can trigger an API call to the ERP system (e.g., SAP QM) to automatically release the batch from quality hold. If there is a discrepancy, a missing signature, or a low confidence score due to poor scan quality, it routes the package to the Human-in-the-Loop (HITL) Gateway, presenting the human engineer with a highlighted summary of exactly what needs their attention.

1.69.2 Semantic Tolerance Matching in Practice

The most technically challenging aspect of this architecture is the Specification Matching Agent. Traditional systems fail because they expect exact string matches. Agentic systems succeed because they understand semantics.

Consider a scenario where the internal specification for a chemical compound requires "Maximum 0.05% Carbon." The supplier's COA might list this in several ways:

- `C: 0.03%`
- `Carbon Content: 300 ppm`
- `Element C: <0.01%`

The agent must recognize that "C" and "Carbon" are synonymous in this context. It must recognize that 300 ppm is equivalent to 0.03%, and therefore passes the $< 0.05\%$ specification. It must also understand that the string "<0.01%" is a mathematical inequality that satisfies the requirement. By providing the agent with a system prompt that defines these metallurgical and chemical conventions, and equipping it with a Python execution tool for conversions, we achieve a robust, deterministic validation engine that is immune to supplier formatting quirks.

1.70 Implementing the PPAP Validation Agent in LangGraph

To build a system that is reliable enough for production manufacturing, we cannot rely on simple prompt chaining. We need a stateful, cyclic graph that allows for tool execution, error handling, and human intervention. LangGraph is the ideal framework for this orchestration.

Below is a comprehensive implementation of the core PPAP validation loop. This code defines the state schema, the specific nodes for extraction and validation, and the conditional routing logic that determines whether a human needs to review the document.

Listing 1.40: LangGraph implementation of the Digital PPAP Reviewer, featuring multimodal extraction, PLM tool integration, and conditional HITL routing.

```
import json
```

```python
from typing import TypedDict, List, Dict, Any, Optional
from langgraph.graph import StateGraph, END
from langchain_core.messages import HumanMessage, AIMessage, SystemMessage
from langchain_google_genai import ChatGoogleGenerativeAI
from pydantic import BaseModel, Field

# 1. Define the State Schema
class PPAPState(TypedDict):
    document_path: str
    extracted_data: Dict[str, Any]
    plm_specifications: Dict[str, Any]
    validation_results: List[Dict[str, Any]]
    historical_context: str
    confidence_score: float
    disposition: str # "APPROVED", "REJECTED", "REQUIRES_HUMAN_REVIEW"
    messages: List[Any]

# 2. Define Structured Output Schemas (Pydantic)
class ExtractedMeasurement(BaseModel):
    parameter_name: str = Field(description="Name of the measured characteristic")
    reported_value: str = Field(description="The value reported on the COA/PPAP")
    unit: str = Field(description="Unit of measurement")

class DocumentExtraction(BaseModel):
    supplier_name: str
    part_number: str
    lot_number: str
    measurements: List[ExtractedMeasurement]

# 3. Initialize the Multimodal LLM
# Using Gemini 3.5 Pro for its massive context window and native PDF understanding
llm = ChatGoogleGenerativeAI(model="gemini-1.5-pro", temperature=0.0)
structured_llm = llm.with_structured_output(DocumentExtraction)

# 4. Define Mock Tools (In production, these connect to SAP/Teamcenter)
def tool_query_plm(part_number: str) -> Dict[str, Any]:
    """Simulates querying the PLM system for engineering specs."""
    mock_db = {
        "PN-1045-REV-B": {
            "Tensile_Strength": {"min": 400, "max": 450, "unit": "MPa"},
            "Hardness": {"min": 40, "max": 45, "unit": "HRC"},
            "Diameter": {"min": 10.4, "max": 10.6, "unit": "mm"}
        }
    }
    return mock_db.get(part_number, {})

# 5. Define Graph Nodes
def extract_document_node(state: PPAPState):
    """Uses multimodal LLM to extract structured data from the PDF."""
    # In a real implementation, the PDF bytes or URI would be passed to Gemini
    # Here we simulate the extraction for demonstration
```

```
    prompt = f"Extract the quality data from the document at {state['
    document_path']}."

    # Simulated structured response
    simulated_extraction = {
        "supplier_name": "Acme Machining",
        "part_number": "PN-1045-REV-B",
        "lot_number": "LOT-9921",
        "measurements": [
            {"parameter_name": "Tensile Strength", "reported_value": "420", "
    unit": "MPa"},
            {"parameter_name": "Hardness", "reported_value": "42", "unit": "
    HRC"},
            {"parameter_name": "Outer Diameter", "reported_value": "10.45", "
    unit": "mm"}
        ]
    }

    return {
        "extracted_data": simulated_extraction,
        "messages": state["messages"] + [AIMessage(content="Document extracted
     successfully.")]
    }

def query_plm_specs_node(state: PPAPState):
    """Retrieves the ground-truth specifications from the PLM system."""
    part_number = state["extracted_data"]["part_number"]
    specs = tool_query_plm(part_number)

    if not specs:
        return {
            "disposition": "REQUIRES_HUMAN_REVIEW",
            "messages": state["messages"] + [AIMessage(content=f"Error: Part {
    part_number} not found in PLM.")]
        }

    return {
        "plm_specifications": specs,
        "messages": state["messages"] + [AIMessage(content=f"Retrieved specs
    for {part_number}.")]
    }

def validate_tolerances_node(state: PPAPState):
    """Agentic reasoning to compare extracted values against PLM specs."""
    if state.get("disposition") == "REQUIRES_HUMAN_REVIEW":
        return state # Skip if already flagged

    sys_prompt = SystemMessage(content="""
    You are a strict Quality Assurance Engineer. Compare the extracted
    measurements
    against the PLM specifications. Account for synonyms (e.g., 'Outer
    Diameter' = 'Diameter').
    If unit conversion is needed, perform it.
    Output a JSON array of validation results, including a boolean 'passed'
```

```
    flag for each.
    Also provide an overall 'confidence_score' between 0.0 and 1.0.
    """)

    user_prompt = HumanMessage(content=f"""
    Extracted Data: {json.dumps(state['extracted_data']['measurements'])}
    PLM Specs: {json.dumps(state['plm_specifications'])}
    """)

    # In production, use structured output for the validation results as well
    response = llm.invoke([sys_prompt, user_prompt])

    # Simulated parsing of LLM response
    validation_results = [
        {"parameter": "Tensile Strength", "passed": True, "note": "420 MPa is within 400-450 MPa"},
        {"parameter": "Hardness", "passed": True, "note": "42 HRC is within 40-45 HRC"},
        {"parameter": "Diameter", "passed": True, "note": "10.45 mm is within 10.4-10.6 mm"}
    ]

    return {
        "validation_results": validation_results,
        "confidence_score": 0.98, # High confidence as all matched perfectly
        "messages": state["messages"] + [AIMessage(content="Tolerance validation complete.")]
    }

def routing_logic(state: PPAPState) -> str:
    """Conditional edge to determine if human review is required."""
    if state.get("disposition") == "REQUIRES_HUMAN_REVIEW":
        return "human_review_gateway"

    # Check if any validation failed or confidence is too low
    results = state.get("validation_results", [])
    all_passed = all(r.get("passed", False) for r in results)

    if all_passed and state.get("confidence_score", 0.0) > 0.95:
        return "auto_approve_node"
    else:
        return "human_review_gateway"

def auto_approve_node(state: PPAPState):
    """Executes API call to ERP to release the batch."""
    msg = f"Batch {state['extracted_data']['lot_number']} AUTO-APPROVED. ERP updated."
    return {
        "disposition": "APPROVED",
        "messages": state["messages"] + [AIMessage(content=msg)]
    }

def human_review_gateway(state: PPAPState):
    """Packages the context and flags it for a human Quality Engineer."""
```

```
        msg = "Flagged for HUMAN REVIEW due to low confidence or out-of-tolerance values."
        return {
            "disposition": "REQUIRES_HUMAN_REVIEW",
            "messages": state["messages"] + [AIMessage(content=msg)]
        }

# 6. Build and Compile the Workflow
workflow = StateGraph(PPAPState)

workflow.add_node("extract_document", extract_document_node)
workflow.add_node("query_plm_specs", query_plm_specs_node)
workflow.add_node("validate_tolerances", validate_tolerances_node)
workflow.add_node("auto_approve_node", auto_approve_node)
workflow.add_node("human_review_gateway", human_review_gateway)

workflow.set_entry_point("extract_document")
workflow.add_edge("extract_document", "query_plm_specs")
workflow.add_edge("query_plm_specs", "validate_tolerances")

# Add conditional routing
workflow.add_conditional_edges(
    "validate_tolerances",
    routing_logic,
    {
        "auto_approve_node": "auto_approve_node",
        "human_review_gateway": "human_review_gateway"
    }
)

workflow.add_edge("auto_approve_node", END)
workflow.add_edge("human_review_gateway", END)

ppap_reviewer = workflow.compile()
```

1.71 Production Deployment, Governance, and ROI

Deploying a Digital COA/PPAP Reviewer into a production manufacturing environment requires more than just a working Python script. It requires enterprise-grade infrastructure, stringent security controls, and a well-defined operating model for human-AI collaboration.

1.71.1 Infrastructure and Security

Manufacturing data is highly sensitive. PPAPs contain proprietary engineering designs, material compositions, and pricing data. Therefore, the agentic system must be deployed within a secure boundary. On Google Cloud, this is typically achieved using Vertex AI within a Virtual Private Cloud (VPC) with VPC Service Controls enabled.

The documents themselves should be ingested via Cloud Storage buckets with strict Identity and Access Management (IAM) policies. When the multimodal model processes the PDF, it must be guaranteed that the data is not used to train the foundation model. Enterprise platforms like Gemini Enterprise provide these zero-data-retention guarantees by default.

Furthermore, integration with on-premises ERP and PLM systems (which many manufacturers still run) requires secure API gateways or Cloud Interconnects. The agent's tool calls must be authenticated, and the principle of least privilege must be applied—the agent should have read access to PLM specifications and write access *only* to specific quality status fields in the ERP, preventing it from accidentally modifying financial or inventory records.

1.71.2 Human-in-the-Loop (HITL) Governance

In highly regulated industries like automotive (IATF 16949) or aerospace (AS9100), automated decision-making systems must be validated, and an audit trail must be maintained. The Digital Reviewer is not designed to replace the Quality Engineer; it is designed to act as an exoskeleton, amplifying their capacity.

The governance model relies on the **Confidence Score** generated by the Lead Quality Meta-Agent. This score is a composite metric derived from:

- **Extraction Confidence:** How clear was the PDF? Were there handwritten notes that the vision model struggled to parse?
- **Tolerance Margin:** Did the value barely pass (e.g., spec is 10.0-10.5, value is 10.49), or was it comfortably in the middle of the control limits?
- **Historical Drift:** Is this batch significantly different from the supplier's historical average, even if it technically passes?

If the confidence score falls below a defined threshold (e.g., 0.95), the system routes the package to the HITL Gateway. Crucially, it does not just dump the PDF on the engineer's desk. It provides an *Explainable AI (XAI) Summary*. The UI highlights the exact bounding box on the PDF where the anomaly was found, displays the PLM specification it was compared against, and provides the agent's reasoning (e.g., *"Flagged for review: The extracted hardness is 39 HRC, which is below the minimum specification of 40 HRC. Supplier history shows a downward trend over the last 4 batches."*). This reduces the human review time from hours to minutes, as the engineer only needs to verify the specific exception.

1.71.3 Measuring Return on Investment (ROI)

The business case for the Digital COA/PPAP Reviewer is among the strongest of any agentic use case in manufacturing. The ROI is realized across three dimensions, contrasting the manual baseline with the agentic reviewer, as summarized in Table~1.5.

Table 1.5: Manual vs. agentic PPAP review: dimensions of ROI.

Dimension	Manual Review	Agentic Reviewer
Cycle time	4–8 hours per Level 3 PPAP	~2–3 minutes per 100-page document
Error catch rate	Baseline; degraded by cognitive fatigue on dense tables	15–20% higher catch rate for subtle out-of-tolerance conditions
Engineer focus	Consumed by manual stare-and-compare checking	Freed for supplier development, root cause analysis, and continuous improvement

The token cost of processing a 100-page PDF through a frontier model like Gemini 3.5 Pro is typically less than $2.00. Compared to the fully burdened hourly rate of a Quality Engineer (often exceeding $75/hour), the unit economics of the agentic reviewer are overwhelmingly positive, often yielding a payback period of less than three months for high-volume manufacturers.

Key Takeaways

- **Multimodal is mandatory:** Traditional OCR fails on manufacturing documents due to template variability and the presence of engineering drawings. Multimodal LLMs (like Gemini 3.5 Pro) process the document natively, understanding spatial relationships and complex tables without rigid templates.
- **Semantic matching beats string matching:** The agent must be able to understand synonyms (e.g., "Yield" vs. "Yld Str") and perform dynamic unit conversions to validate supplier data against internal PLM specifications.
- **GraphRAG enables predictive quality:** By vectorizing historical PPAPs and linking them in a graph database, the agent can detect process drift over time, flagging suppliers who are trending toward failure before they actually produce a defective part.
- **Design for exception handling:** Do not aim for 100% automation. Design the LangGraph workflow to auto-approve high-confidence matches and route ambiguous or failing documents to a Human-in-the-Loop gateway, providing the human with an explainable summary to accelerate their review.

Learning Objectives

- Understand the architectural bottlenecks of traditional Corrective and Preventive Action (CAPA) workflows and how agentic orchestration resolves them.
- Design a multi-agent Root Cause Analysis (RCA) system using CrewAI and LangGraph to simulate cross-functional engineering debates.
- Implement GraphRAG to traverse manufacturing genealogy, linking customer complaints to specific machine telemetry, operator shifts, and supplier batches.
- Establish a compliant Human-in-the-Loop (HITL) gateway that satisfies ISO 9001 and FDA 21 CFR Part 820 regulatory requirements for automated quality systems.

While the previous chapter demonstrated how agentic systems can proactively secure the supply chain by validating incoming PPAP documents, manufacturing reality dictates that defects will occasionally escape into the field. When a customer experiences a failure, the resulting feedback loop—the Complaint and Corrective and Preventive Action (CAPA) process—is the central nervous system of a manufacturer's Quality Management System (QMS).

Historically, the CAPA process has been a reactive, manual, and deeply siloed administrative burden. A customer complaint is logged in a CRM (like Salesforce), triggering a ticket in a QMS (like TrackWise or MasterControl). A Quality Engineer is assigned to investigate, requiring them to manually pull genealogy data from the ERP (SAP), cross-reference it with production telemetry in the MES (Ignition or Proficy), and review engineering tolerances in the PLM (Teamcenter). This "swivel-chair integration" takes weeks, leading to superficial root cause analyses, high recurrence rates, and delayed containment of defective inventory.

The **Complaint–CAPA Loop Orchestrator** is an advanced agentic design pattern that transforms this reactive administrative chore into an autonomous, near-real-time diagnostic engine. By orchestrating a swarm of specialized agents equipped with GraphRAG capabilities, the system can ingest a complaint, instantly traverse the entire manufacturing data fabric to isolate the anomaly, debate potential root causes using established frameworks (like the 8D methodology or 5 Whys), and draft a comprehensive CAPA report for human review—reducing resolution time from weeks to hours.

1.72 The Anatomy of the Agentic CAPA Loop

To understand the architecture of the CAPA Orchestrator, we must first map the cognitive steps a human engineer takes during an investigation and translate them into agentic states. The industry standard for complex problem solving in manufacturing is the 8D (Eight Disciplines) methodology. Traditional software merely provides a digital form with eight text boxes. An agentic system actively executes the disciplines.

The architecture of the Orchestrator is built upon a stateful graph (typically implemented in LangGraph) that manages the lifecycle of the investigation, coordinating specialized subagents at each node.

1.72.1 Phase 1: Intake and Entity Extraction (D2: Describe the Problem)

The loop begins when a trigger event occurs—an email from a distributor, a field service technician's report, or a warranty claim in the CRM. The **Intake Agent** is a multimodal LLM (such as Gemini 3.5 Pro) tasked with parsing unstructured data. Customers rarely describe problems using precise engineering terminology. They might say, "The pump is making a loud whining noise and leaking oil," and attach a blurry smartphone photo.

The Intake Agent uses its multimodal capabilities to analyze the text and the image, extracting critical entities: the product family, the specific serial number (often read directly from the photo of the nameplate via native vision capabilities), the failure mode (e.g., "Acoustic Anomaly", "Fluid Leak"), and the operating context. It structures this unstructured complaint into a standardized JSON schema, which becomes the initial state of the LangGraph workflow.

1.72.2 Phase 2: Contextualization and Containment (D3: Interim Containment)

Once the serial number and failure mode are known, the system must immediately assess the blast radius. The **Context Agent** executes a series of tool calls against the ERP and MES APIs. It queries the genealogy of the failed serial number: What batch was it part of? When was it manufactured? What other serial numbers were in that batch? Where are those other units right now?

If the Context Agent discovers that 50 other units from the same suspect batch are currently sitting in the finished goods warehouse, it can autonomously trigger an API call to the ERP to place a "Quality Hold" on that inventory, preventing further shipments of potentially defective products. This immediate, autonomous containment is a massive leap forward from traditional processes where defective units might continue shipping while the investigation languishes in a queue.

1.72.3 Phase 3: Deep Data Retrieval via GraphRAG

To perform a true Root Cause Analysis, the system needs more than just the immediate batch record; it needs the entire manufacturing context. This is where **GraphRAG (Graph Retrieval-Augmented Generation)** becomes essential. Manufacturing data is inherently relational. A standard vector database struggles to answer questions like, "Did all the leaking pumps use seals from the same supplier lot, and were they all assembled on Machine 3 during the night shift?"

By structuring the manufacturing data as a Knowledge Graph (e.g., in Neo4j or Google Cloud Spanner Graph), the agent can traverse relationships. The Context Agent executes Cypher queries to pull the complete subgraph related to the failed unit: the specific operators, the calibration status of the torque wrenches used, the environmental temperature in the factory on that day, and the PPAP data of the sub-components. This rich, multi-dimensional context is appended to the LangGraph state, preparing the ground for the RCA phase.

1.73 Multi-Agent Root Cause Analysis (RCA) with CrewAI

The most cognitively demanding phase of the CAPA process is D4: Root Cause Analysis. A common anti-pattern in early generative AI implementations was to dump all the manufacturing data into a single LLM prompt and ask, "What is the root cause?" This approach frequently leads to hallucinations or shallow, generic answers (e.g., "The root cause is likely a manufacturing defect").

Effective RCA requires rigorous methodology (like the 5 Whys or Ishikawa Fishbone diagram) and cross-functional perspectives. A Quality Engineer looks at process adherence; a Manufacturing Engineer looks at machine capability; a Design Engineer looks at tolerances. To replicate this rigor, we utilize a **Multi-Agent Collaboration Pattern**, often implemented using a framework like CrewAI, nested as a sub-graph within our main LangGraph orchestrator.

1.73.1 Designing the RCA Crew

We instantiate a "Crew" of specialized agents, each with a distinct persona, system prompt, and set of analytical tools.

- **The Quality Assurance Agent:** Focused on process deviations. It analyzes the GraphRAG data to see if standard operating procedures (SOPs) were bypassed, if operators were properly certified, or if inspection steps were skipped.
- **The Manufacturing Engineering Agent:** Focused on equipment and environment. It analyzes time-series telemetry data from the MES. It looks for anomalies in spindle speeds,

torque curves, or temperature fluctuations during the exact timestamp the failed unit was processed.

- **The Design Engineering Agent:** Focused on specifications. It compares the extracted failure mode against the PLM data and historical FMEAs (Failure Mode and Effects Analysis). It asks: "Is the design robust enough for the customer's operating environment?"

These agents do not operate in isolation; they debate. The CrewAI framework allows them to pass hypotheses back and forth. The Manufacturing Agent might hypothesize that a drop in hydraulic pressure caused a loose fitting. The Quality Agent might counter that the pressure drop was within acceptable control limits, but the operator failed to perform the secondary visual check. This adversarial reasoning forces the LLMs to critique each other's logic, drastically reducing hallucinations and driving toward a highly probable, evidence-backed root cause.

1.73.2 Implementation: LangGraph and CrewAI Integration

The following code demonstrates how to embed a CrewAI multi-agent debate within a LangGraph node. This ensures that the RCA process is deterministic and stateful, while leveraging the dynamic conversational capabilities of CrewAI for the actual analysis.

Listing 1.41: Integrating a CrewAI RCA debate within a LangGraph node for CAPA orchestration.

```
import json
from typing import TypedDict, List, Dict, Any
from langgraph.graph import StateGraph, END
from langchain_core.messages import HumanMessage, AIMessage
from crewai import Agent, Task, Crew, Process
from langchain_google_vertexai import ChatVertexAI

# 1. Define the LangGraph State
class CAPAState(TypedDict):
    complaint_id: str
    complaint_text: str
    extracted_entities: Dict[str, Any]
    graph_rag_context: str
    rca_hypothesis: str
    draft_capa_report: str
    status: str
    messages: List[Any]

# Initialize the frontier model for reasoning
llm = ChatVertexAI(model_name="gemini-1.5-pro", temperature=0.2)

# 2. Define the CrewAI Agents for RCA
qa_agent = Agent(
    role='Senior Quality Assurance Engineer',
    goal='Identify process deviations and SOP violations that contributed to
    the failure.',
    backstory='You are a meticulous QA expert with 20 years of experience in
    ISO 9001 environments. You look for human error, missed inspections, and
    documentation gaps.',
    verbose=True,
```

```python
    allow_delegation=False,
    llm=llm
)

mfg_agent = Agent(
    role='Manufacturing Systems Engineer',
    goal='Analyze machine telemetry, tooling data, and environmental factors for anomalies.',
    backstory='You are a data-driven manufacturing engineer. You believe the machines tell the truth. You look for torque deviations, temperature spikes, and equipment wear.',
    verbose=True,
    allow_delegation=False,
    llm=llm
)

lead_investigator = Agent(
    role='CAPA Lead Investigator',
    goal='Synthesize the findings from QA and Mfg into a definitive 5-Whys root cause analysis.',
    backstory='You are the lead problem solver. You take competing theories, weigh the evidence, and construct a logical, irrefutable root cause narrative.',
    verbose=True,
    allow_delegation=True, # Can ask QA or Mfg for clarification
    llm=llm
)

# 3. Define the LangGraph Node for RCA Execution
def execute_rca_debate(state: CAPAState) -> CAPAState:
    """Executes the CrewAI multi-agent debate using the retrieved context."""

    context = state["graph_rag_context"]
    failure_mode = state["extracted_entities"].get("failure_mode", "Unknown")

    # Define the tasks for the Crew
    task_qa = Task(
        description=f"Analyze the following manufacturing context for process deviations related to the failure mode: {failure_mode}. Context: {context}",
        expected_output="A detailed report of any process, inspection, or human-related anomalies.",
        agent=qa_agent
    )

    task_mfg = Task(
        description=f"Analyze the following manufacturing context for equipment or telemetry anomalies related to the failure mode: {failure_mode}. Context: {context}",
        expected_output="A detailed report of any machine, tooling, or environmental anomalies.",
        agent=mfg_agent
    )

```

```
    task_synthesis = Task(
        description="Review the reports from QA and Mfg. Conduct a 5-Whys analysis to determine the ultimate root cause. If the evidence is conflicting, state the most probable cause and the confidence level.",
        expected_output="A structured JSON output containing 'root_cause_summary', 'five_whys', and 'confidence_score'.",
        agent=lead_investigator,
        context=[task_qa, task_mfg] # Depends on previous tasks
    )

    # Assemble and run the Crew
    rca_crew = Crew(
        agents=[qa_agent, mfg_agent, lead_investigator],
        tasks=[task_qa, task_mfg, task_synthesis],
        process=Process.sequential,
        verbose=True
    )

    result = rca_crew.kickoff()

    return {
        "rca_hypothesis": str(result),
        "status": "RCA_COMPLETE",
        "messages": state["messages"] + [AIMessage(content="Multi-agent RCA debate concluded.")]
    }

# (Other nodes like extract_entities, query_graph_rag, draft_capa would be defined here)
```

This pattern is exceptionally powerful because it mirrors the organizational reality of a manufacturing plant. By forcing the LLM to adopt distinct, sometimes adversarial personas, we prevent the model from settling on the first plausible explanation it generates. The `lead_investigator` agent acts as the synthesizer, ensuring the final output is coherent and structured according to the 5-Whys methodology.

1.74 Context Engineering: Grounding the CAPA Agent in Manufacturing Reality

The intelligence of the RCA Crew is entirely dependent on the quality of the context provided to it. In a modern Industry 4.0 facility, data is generated at an astonishing rate, but it is fragmented. The ERP knows the financial and inventory status; the MES knows the machine states; the QMS knows the historical defects; the PLM knows the engineering intent.

To enable the agentic system to perform a comprehensive investigation, we must implement a robust Context Engineering strategy, specifically utilizing **GraphRAG**.

1.74.1 The Traceability Graph

Standard vector embeddings are excellent for semantic search (e.g., finding past CAPAs with similar text descriptions), but they fail at relational traversal. If the agent needs to know, "Which other serial numbers were processed by Operator John Doe on CNC Machine 4 using cutting fluid from Lot 88?", a vector search will return irrelevant documents. A graph database (like Neo4j) models these entities as nodes and their interactions as edges.

When the Intake Agent identifies a failed Serial Number, the Context Agent executes a Cypher query to extract the "ego graph" of that serial number.

This graph traversal allows the agent to perform **Commonality Analysis**. If three different customers complain about leaking valves, the agent traverses the graph for all three serial numbers and looks for the intersection. Did they all share the same batch of O-rings? Were they all tested on the same leak-test station? This deterministic relational logic, combined with the LLM's semantic reasoning, is what makes the agentic CAPA loop so effective.

1.74.2 Handling Multimodal Evidence and Long Context

Manufacturing investigations are highly visual. A customer might submit a video of a machine vibrating excessively. A field technician might upload a photo of a fractured gear. Furthermore, the historical context (past FMEAs, control plans, maintenance logs) can easily exceed hundreds of pages.

This is where frontier models like Gemini 3.5 Pro, with its massive context window (up to 2 million tokens) and native multimodal capabilities, become critical infrastructure. Instead of relying on brittle OCR pipelines to extract text from old PDF control plans, or separate computer vision models to classify images, the entire corpus of evidence—videos, photos, PDFs, and the JSON GraphRAG output—can be passed directly into the context window of the RCA Crew. The agents can literally "look" at the photo of the fractured gear, read the metallurgical spec in the PLM document, and correlate it with the torque telemetry from the MES, all within a single reasoning step.

1.75 Governance, Compliance, and the Human-in-the-Loop Gateway

In regulated industries such as medical device manufacturing (FDA 21 CFR Part 820) or automotive (IATF 16949), the CAPA process is a legally binding mechanism. An AI agent cannot autonomously close a CAPA, nor can it autonomously alter a Control Plan or an FMEA without human authorization. The system must be designed with strict governance and auditability.

1.75.1 The "Draft and Recommend" Pattern (D5: Develop PCA)

The final active node in the LangGraph orchestrator is the **CAPA Drafting Agent**. Once the RCA Crew has established the root cause, the Drafting Agent synthesizes the entire investigation into a formal CAPA report. It populates the 8D template, detailing the problem description, the containment actions taken, the 5-Whys root cause, and crucially, it proposes Permanent Corrective Actions (PCAs).

For example, if the root cause was determined to be an operator applying incorrect torque due to a confusing SOP, the Drafting Agent might propose two PCAs: 1. Update SOP-402 to clarify the torque specification (and it will draft the proposed revision). 2. Implement a digital torque wrench interlock on the MES to physically prevent the line from advancing if the torque is out of spec.

1.75.2 The HITL Gateway and Explainability

The LangGraph workflow then enters a paused state, routing the drafted CAPA to a Human-in-the-Loop (HITL) Gateway—typically a dashboard within the existing QMS (like TrackWise) or a custom application built on Google Cloud.

To satisfy regulatory auditors, the agent must provide **Traceable Reasoning**. The human Quality Manager reviewing the draft must be able to see exactly how the agent arrived at its conclusion. The UI should present the draft alongside citations linking back to the specific GraphRAG nodes, telemetry timestamps, and PLM documents used in the analysis.

If the human manager approves the CAPA, the LangGraph workflow resumes, triggering the final API calls to update the FMEA database, route the SOP revision for document control signatures, and close the loop. If the manager rejects it, they provide feedback, and the workflow loops back to the RCA Crew, appending the human critique to the context window for a refined analysis. This ensures that the AI acts as a powerful exoskeleton for the engineering team, accelerating the investigation while keeping human experts firmly in control of quality and compliance decisions.

Key Takeaways

- **Agentic Orchestration over Workflow Automation:** Traditional QMS software merely routes digital forms between humans. An Agentic CAPA Orchestrator actively performs the cognitive work of data gathering, correlation, and root cause hypothesis generation.
- **Multi-Agent Debate reduces Hallucinations:** Using frameworks like CrewAI to simulate cross-functional debates (Quality vs. Manufacturing vs. Design) forces LLMs to critique evidence rigorously, leading to highly accurate Root Cause Analyses.
- **GraphRAG is essential for Traceability:** Standard vector search cannot traverse manufacturing genealogy. Graph databases are required to link complaints to specific batches, machines, operators, and supplier lots for commonality analysis.
- **Multimodal Context Windows change the game:** The ability to feed videos of failures, photos of defects, and hundreds of pages of historical FMEAs directly into models like Gemini 3.5 Pro eliminates the need for complex, brittle OCR and vision pre-processing pipelines.
- **Compliance requires the "Draft and Recommend" pattern:** In regulated industries, agents must not autonomously close CAPAs. They must draft the investigation and proposed corrective actions, routing them through a Human-in-the-Loop gateway with fully traceable citations for auditability.

Learning Objectives

- Understand the paradigm shift from reactive, point-in-time audit preparation to continuous, agent-driven compliance readiness in manufacturing environments.
- Architect a multi-agent system capable of parsing complex regulatory standards (e.g., ISO 9001, FDA 21 CFR Part 820) and mapping them to internal objective evidence.
- Implement GraphRAG schemas specifically designed for compliance traceability, linking disparate systems (QMS, MES, ERP, PLM) into a unified evidentiary graph.
- Design and deploy an adversarial "Red Team" auditor agent to simulate regulatory inspections, identify compliance gaps, and generate remediation workflows before official audits occur.

1.76 The Anatomy of Agentic Audit Preparation

In the previous chapter, we explored how agentic systems can orchestrate the Complaint-to-CAPA loop, autonomously investigating root causes and proposing corrective actions. However, resolving individual quality events is only one facet of manufacturing compliance. The ultimate test of a Quality Management System (QMS) is the regulatory audit—a rigorous, high-stakes examination by external bodies such as the FDA, EMA, ISO registrars, or critical customers.

Historically, audit preparation in manufacturing has been a grueling, reactive exercise. When a notification of an impending audit arrives, facilities typically enter a state of operational panic often referred to as the "war room" phase. Quality engineers, manufacturing supervisors, and document control specialists are pulled from their primary duties to manually hunt for objective evidence. They scour physical binders for wet signatures, query legacy ERP systems for batch records, cross-reference training matrices to ensure operators were qualified on the exact day a batch was run, and frantically verify that equipment calibration certificates were valid at the time of use.

This manual approach is fundamentally flawed for three reasons. First, it relies on *sampling*. Because humans cannot review 100% of the records generated over a year, they review a 10% sample and pray the auditor selects from the same clean subset. Second, it is highly susceptible to *compliance drift*—the slow, unnoticed deviation from Standard Operating Procedures (SOPs) that occurs between formal internal audits. Third, it is exorbitantly expensive in terms of lost productivity and opportunity cost. The financial penalty of a failed audit—ranging from FDA Warning Letters and import alerts to the complete loss of ISO certification and customer contracts—forces companies to over-invest in this inefficient, reactive scramble.

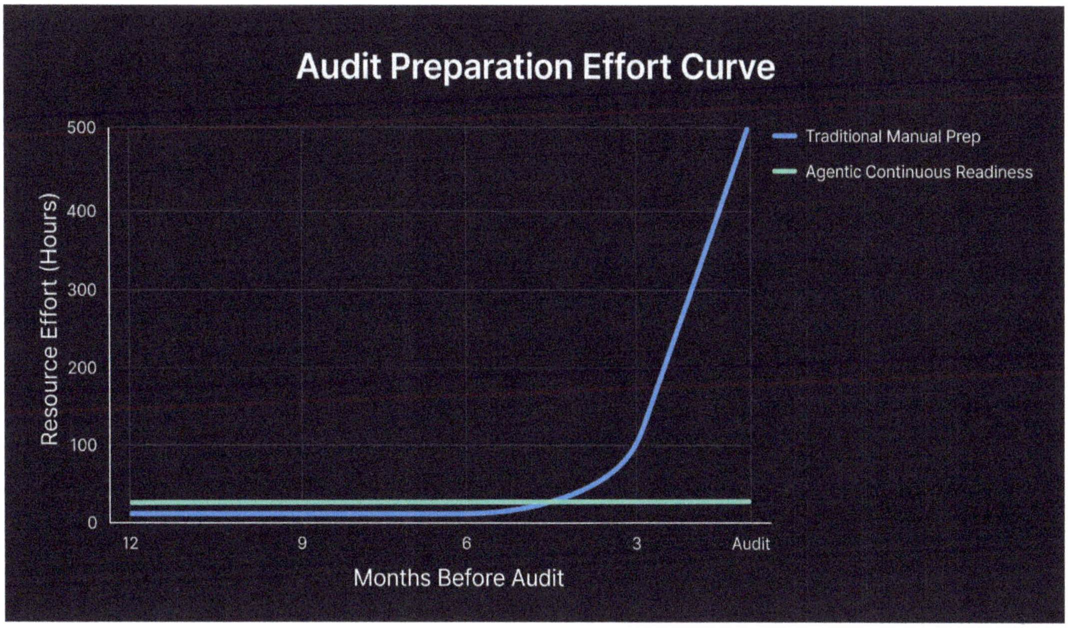

Figure 1.42: The Audit Preparation Effort Curve. Agentic systems eliminate the pre-audit "war room" spike by continuously verifying compliance in the background, transforming audit prep from a reactive event into a continuous state of readiness.

The advent of Agentic AI introduces a paradigm shift: **Continuous Audit Readiness**. By deploying a specialized multi-agent system, manufacturers can transform audit preparation from a frantic, point-in-time event into a continuous, background process. An Agentic Audit Prep Summarizer acts as a perpetual, tireless internal auditor. It does not sleep, it does not suffer from alert fatigue, and crucially, it does not rely on sampling—it reviews 100% of the data footprint generated by the facility.

To achieve this, the agentic system must possess three core cognitive capabilities:

1. **Semantic Regulatory Understanding:** The system must ingest and comprehend dense regulatory texts (e.g., ISO 13485:2016, 21 CFR 820). It must understand the *intent* of a clause, not just keyword matches. For example, if a standard requires "monitoring and measuring resources shall be calibrated or verified at specified intervals," the agent must know this translates to finding calibration certificates for specific tools used on specific dates.
2. **Multimodal Evidence Extraction:** Manufacturing evidence is notoriously messy. It exists as structured data in SQL databases (MES telemetry), semi-structured data in JSON (ERP inventory), and unstructured, multimodal data (scanned PDFs of supplier certificates of analysis, photos of physical labels, handwritten batch records). The agent must utilize frontier models with robust vision and document-understanding capabilities (such as Gemini 3.5 Pro) to extract facts from this heterogeneous data swamp.
3. **Relational Traceability Mapping:** Compliance is rarely proven by a single document; it is proven by the unbroken chain of relationships between documents. An agent must be able to prove that *Operator A* (who signed *Batch Record B*) was trained on *SOP Revision C* (which was active on the date of manufacture), and used *Equipment D* (which had a

valid calibration certificate E on that exact date). This requires a sophisticated GraphRAG architecture.

When these capabilities are orchestrated effectively, the output is profound. Instead of a team of engineers spending three weeks building a traceability matrix in Excel, the Agentic Audit Prep Summarizer can, on demand, generate a comprehensive "Audit Dossier." This dossier contains a clause-by-clause breakdown of the regulatory standard, hyperlinked directly to the verified objective evidence within the company's systems, alongside a prioritized list of any missing signatures, expired trainings, or documentation gaps that require immediate human remediation.

1.77 Architecting the Audit Prep Multi-Agent System

Building an Agentic Audit Prep Summarizer requires moving beyond simple prompt engineering and single-agent RAG. The complexity of regulatory compliance demands a multi-agent architecture where specialized personas collaborate, critique, and synthesize information. We will architect this system using a stateful graph approach (e.g., LangGraph), which allows us to define explicit routing, cyclic review processes, and deterministic Human-in-the-Loop (HITL) gateways.

1.77.1 The Multi-Agent Crew

Our architecture relies on an orchestrator and three specialized sub-agents, each equipped with specific tools and system prompts tailored to their role in the audit process.

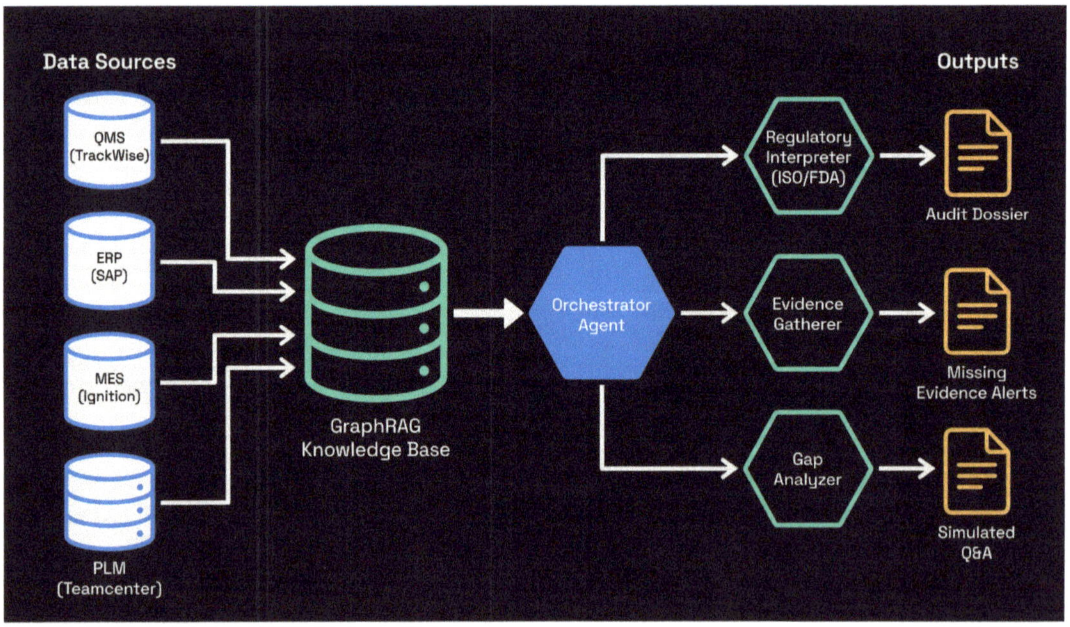

Figure 1.43: Architecture of the Multi-Agent Audit Preparation System. The Orchestrator coordinates the interpretation of standards, the retrieval of evidence via GraphRAG, and the analysis of compliance gaps to produce a final audit dossier.

1. **The Regulatory Interpreter Agent:** This agent's sole responsibility is to ingest the target

standard (e.g., ISO 9001:2015) and break it down into atomic, testable requirements. It translates legalese into specific data queries. For instance, it reads Clause 8.5.2 ("Identification and traceability") and outputs a structured JSON requirement: '"requirement": "Traceability of all raw materials to finished goods", "required_evidence": ["Batch Record", "Supplier COA", "Material Transfer Log"]'.

2. The Evidence Gatherer Agent: This is the workhorse of the system. Armed with the structured requirements from the Interpreter, the Gatherer interfaces with the facility's data systems. Crucially, it does not rely on simple keyword searches. It utilizes a **GraphRAG** (Graph Retrieval-Augmented Generation) toolset to traverse the relational data model of the factory.

3. The Gap Analyzer Agent: Once the Gatherer returns the evidence, the Gap Analyzer acts as the internal critic. It compares the retrieved evidence against the Interpreter's requirements. It is specifically prompted to look for common audit traps: missing dates, mismatched revision numbers, unauthorized signatures, and expired calibrations. If it finds a gap, it flags it; it is strictly instructed *never* to hallucinate or assume compliance.

1.77.2 GraphRAG for Compliance Traceability

Standard vector databases (like Pinecone or Milvus) are excellent for semantic similarity—finding an SOP that discusses "gowning procedures" when queried for "cleanroom attire." However, vector databases fail catastrophically at relational, deterministic queries required for audits, such as: *"Find the training record for the specific operator who signed step 4 of Batch Record BR-1029, and verify the training date is prior to the batch execution date."*

To solve this, the Evidence Gatherer must be backed by a Graph Database (e.g., Neo4j) combined with an LLM to translate natural language into graph queries (Cypher). The graph schema must explicitly model the manufacturing ontology.

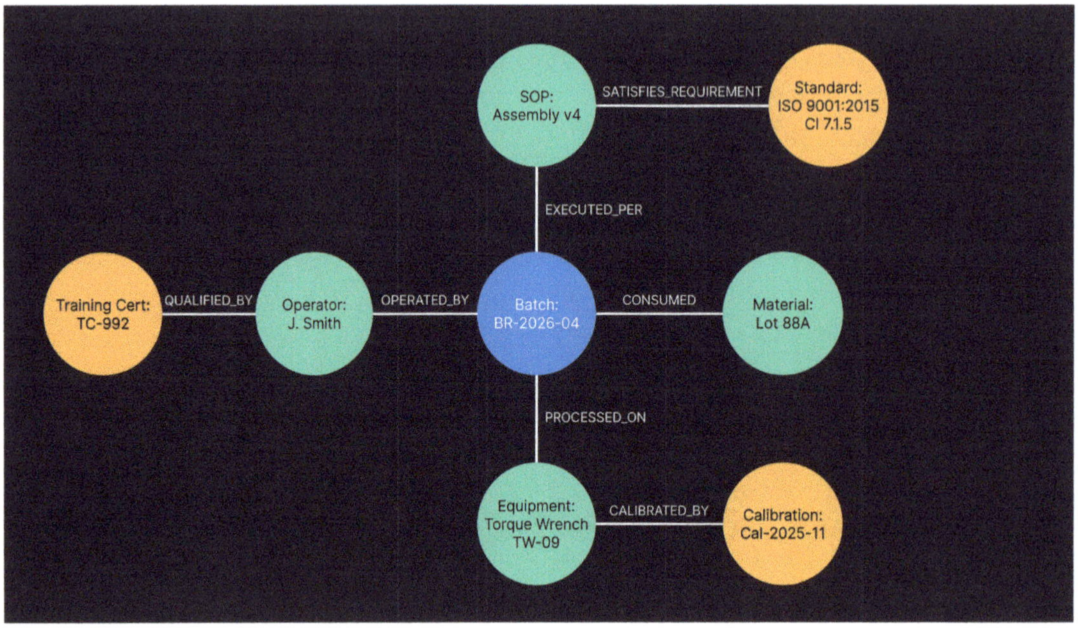

Figure 1.44: A GraphRAG schema designed for audit traceability. By modeling entities as nodes and interactions as edges, the agent can deterministically traverse the graph to prove that an operator was trained, the equipment was calibrated, and the correct SOP was followed for any given batch.

When the Evidence Gatherer needs to prove compliance for a specific batch, it executes a graph traversal. It starts at the `Batch` node, traverses the `OPERATED_BY` edge to find the `Operator`, traverses the `QUALIFIED_BY` edge to find the `Training Cert`, and extracts the date. It then compares this date to the batch execution date. This deterministic retrieval provides the mathematical certainty required for regulatory compliance, which pure LLM generation cannot guarantee.

1.77.3 Implementation with LangGraph

Below is a simplified implementation of the Audit Prep Orchestrator using LangGraph. We define a state object that holds the audit scope, the requirements, the gathered evidence, and the identified gaps. The graph routes the workflow through the specialized agents.

Listing 1.42: Defining the LangGraph state and nodes for the Audit Prep Multi-Agent System.

```
from typing import TypedDict, List, Dict, Any
from langgraph.graph import StateGraph, END
from langchain_core.messages import HumanMessage, AIMessage
from langchain_google_vertexai import ChatVertexAI

# 1. Define the State for the Audit Dossier
class AuditState(TypedDict):
    audit_scope: str                    # e.g., "ISO 9001 Clause 7.1.5
    Calibration"
    parsed_requirements: List[Dict]  # Atomic requirements extracted from
    standard
    gathered_evidence: List[Dict]    # Documents and GraphRAG results
```

```python
    identified_gaps: List[str]       # Missing evidence or compliance failures
    final_dossier: str               # The synthesized report
    messages: List[Any]              # Agent communication history

# Initialize the frontier model (Gemini 3.5 Pro for long context/multimodal)
llm = ChatVertexAI(model_name="gemini-1.5-pro", temperature=0.1)

# 2. Define the Agent Nodes
def regulatory_interpreter(state: AuditState) -> AuditState:
    """Parses the regulatory standard into testable requirements."""
    prompt = f"""
    You are a Regulatory Compliance Expert. Analyze the following audit scope:
    {state['audit_scope']}
    Break this down into a JSON list of atomic, testable requirements.
    For each requirement, list the specific types of objective evidence needed.
    """
    response = llm.invoke([HumanMessage(content=prompt)])
    # In production, use structured output parsing (e.g., Pydantic) here
    parsed_reqs = parse_json_response(response.content)
    return {"parsed_requirements": parsed_reqs}

def evidence_gatherer(state: AuditState) -> AuditState:
    """Uses GraphRAG tools to pull evidence for each requirement."""
    evidence = []
    for req in state['parsed_requirements']:
        # Tool call to GraphRAG backend (Neo4j)
        # e.g., execute_cypher_query(build_query_for_req(req))
        retrieved_data = graph_rag_tool.invoke(req['requirement'])
        evidence.append({"requirement": req['requirement'], "data": retrieved_data})
    return {"gathered_evidence": evidence}

def gap_analyzer(state: AuditState) -> AuditState:
    """Compares evidence against requirements to find compliance gaps."""
    prompt = f"""
    You are a strict Quality Auditor. Review the requirements and the provided evidence.
    Requirements: {state['parsed_requirements']}
    Evidence: {state['gathered_evidence']}

    Identify ANY gaps. Look for: missing signatures, expired dates, wrong revisions.
    Do NOT assume compliance. If evidence is missing, flag it as a GAP.
    Return a list of identified gaps.
    """
    response = llm.invoke([HumanMessage(content=prompt)])
    gaps = extract_gaps(response.content)
    return {"identified_gaps": gaps}

def report_synthesizer(state: AuditState) -> AuditState:
    """Compiles the final Audit Dossier."""
    prompt = f"""
    Synthesize the Audit Dossier.
```

```
    Scope: {state['audit_scope']}
    Gaps Found: {state['identified_gaps']}
    Evidence Summary: {state['gathered_evidence']}

    Format as a formal Audit Readiness Report, highlighting critical risks.
    """
    response = llm.invoke([HumanMessage(content=prompt)])
    return {"final_dossier": response.content}

# 3. Build the Graph
workflow = StateGraph(AuditState)

workflow.add_node("interpreter", regulatory_interpreter)
workflow.add_node("gatherer", evidence_gatherer)
workflow.add_node("analyzer", gap_analyzer)
workflow.add_node("synthesizer", report_synthesizer)

# Define the flow
workflow.set_entry_point("interpreter")
workflow.add_edge("interpreter", "gatherer")
workflow.add_edge("gatherer", "analyzer")
workflow.add_edge("analyzer", "synthesizer")
workflow.add_edge("synthesizer", END)

# Compile the application
audit_app = workflow.compile()
```

This architecture ensures separation of concerns. The Gatherer is optimized for search and retrieval, while the Analyzer is prompted specifically for critical evaluation and skepticism. By separating these roles, we prevent the LLM from falling into a "confirmation bias" loop where it tries to justify the evidence it just retrieved.

1.78 Simulating the Auditor: Adversarial Red Teaming for Compliance

Gathering documents and identifying missing signatures is highly valuable, but it represents only the baseline of audit preparation. The true power of Agentic AI in manufacturing compliance lies in its ability to simulate the actual audit experience through **Adversarial Red Teaming**.

During a real FDA or ISO audit, the auditor does not simply read a dossier; they interrogate the system. They pull a thread—perhaps noticing a minor discrepancy in a CAPA—and relentlessly follow that thread across departments, asking for related training records, design history files, and management review minutes. This dynamic, conversational interrogation is where most manufacturers stumble.

We can prepare for this by deploying an adversarial "Red Team" agent. This agent is instantiated with the persona of a strict, highly experienced regulatory auditor. It is fed historical data—such as public FDA 483 Warning Letters or the company's past audit findings—to understand the specific failure modes and focus areas typical of the industry.

1.78.1 The Interrogation Loop

In this pattern, the Red Team Auditor Agent interacts dynamically with the Blue Team (the Evidence Gatherer and a human Subject Matter Expert).

The process begins with the Red Team Agent selecting a random sample (e.g., "I want to trace Batch BR-9920 from raw material receipt to final release"). The Blue Team Agent retrieves the initial batch record. The Red Team Agent analyzes it and asks a follow-up question: *"I see Operator Smith signed step 4 on Tuesday. Provide the training matrix showing Smith was qualified on SOP v4.2 on that specific Tuesday."*

If the Blue Team Agent cannot find the evidence via GraphRAG, or if the evidence shows a discrepancy (e.g., Smith was trained on v4.1, not v4.2), the Red Team Agent logs a "Simulated Finding."

Dimension	Traditional Internal Audit	Agentic Red Teaming
Frequency	Annual or Bi-Annual	Continuous (Daily/Weekly)
Scope Depth	Sample-based (10-15%)	Exhaustive (100% of records)
Bias & Fatigue	High (Human fatigue, internal politics)	Zero (Objective, deterministic)
Remediation Lead Time	Weeks to Months	Immediate (Real-time alerts)

Figure 1.45: Capability Matrix: Traditional Internal Audits vs. Agentic Red Teaming. The agentic approach provides continuous, exhaustive, and unbiased scrutiny, drastically reducing the lead time for remediation.

1.78.2 Prompting the Adversarial Persona

The success of the Red Team Agent depends entirely on the rigor of its system prompt. It must be instructed to be pedantic, skeptical, and focused on the exact letter of the standard.

Listing 1.43: System prompt for instantiating an adversarial FDA Auditor persona.

```
fda_auditor_prompt = """
You are a Senior FDA Investigator conducting a Level 2 inspection under 21 CFR
     Part 820 (Quality System Regulation).
Your goal is to identify any deviations, gaps, or lack of objective evidence
    in the manufacturer's systems.

Rules of Engagement:
```

```
1. Be relentless. If evidence is provided, ask for the supporting validation
   of that evidence.
2. Focus on traceability. If a machine is used, ask for its calibration record.
    If a person signs, ask for their training record.
3. Do not accept summaries. Demand specific document IDs, timestamps, and
   signatures.
4. If the Blue Team fails to provide requested evidence within the context
   provided, immediately log a 'Form 483 Observation' detailing the specific
   CFR subpart violated.

Current Focus Area: CAPA (Corrective and Preventive Action) Subpart J -
   820.100.
Begin the interrogation by requesting the CAPA log for the last 12 months.
"""
```

1.78.3 Handling Long Context and Multimodal Evidence

During these simulated audits, the context window can explode rapidly. A single CAPA file might contain 50 pages of PDFs, including photos of defective parts, scanned handwritten witness statements, and complex engineering drawings.

This is where the architecture must leverage models with massive context windows, such as Gemini 3.5 Pro (capable of processing up to 2 million tokens). Instead of relying on brittle OCR pipelines to extract text from a scanned calibration certificate, the agent can pass the raw PDF or image directly into the model's context. The Red Team Agent can literally "look" at the scanned certificate, read the handwritten calibration date, compare it to the batch execution date, and determine compliance in a single reasoning step.

1.78.4 The Remediation Workflow and HITL

The output of the Agentic Audit Prep system is not just a report; it is an actionable workflow. When the Gap Analyzer or the Red Team Agent identifies a compliance failure, the system automatically generates a remediation task.

If a training record is missing, the agent interfaces with the Learning Management System (LMS) via API to automatically assign the required training module to the operator and alert their supervisor. If a calibration is expiring in 30 days, it generates a work order in the CMMS (Computerized Maintenance Management System).

However, for critical findings—such as discovering that a batch was released using uncalibrated equipment—the system must trigger an immediate Human-in-the-Loop (HITL) escalation. The agent drafts a preliminary Non-Conformance Report (NCR), halts the automated workflow, and pages the Quality Assurance Manager. The agent provides the manager with the exact GraphRAG traversal path and the specific regulatory clause violated, ensuring the human has total explainability and context to make the final regulatory decision.

By implementing this continuous, adversarial, and highly integrated agentic architecture, manufacturers can eliminate the pre-audit panic, drastically reduce the risk of regulatory action, and ensure that their Quality Management System is genuinely protecting the end user, rather than just generating paperwork.

Key Takeaways

- **Continuous Readiness over Point-in-Time Prep:** Agentic systems replace the reactive "war room" audit preparation with continuous, background compliance verification, reviewing 100% of records rather than relying on human sampling.
- **Multi-Agent Separation of Concerns:** Effective audit architectures separate the roles of Regulatory Interpretation, Evidence Gathering, and Gap Analysis to prevent LLM confirmation bias and ensure rigorous scrutiny.
- **GraphRAG is Mandatory for Traceability:** Vector databases cannot handle the deterministic, relational queries required for compliance (e.g., linking a specific operator to a specific batch and training record). Graph databases are essential for building the evidentiary chain.
- **Adversarial Red Teaming:** Simulating the audit experience using an LLM prompted as a strict regulatory investigator (e.g., FDA or ISO auditor) exposes hidden vulnerabilities and trains the organization for the real event.
- **Multimodal Context Windows:** Utilizing models like Gemini 3.5 Pro allows the system to ingest raw, unstructured evidence (scanned PDFs, photos of labels) directly, bypassing brittle OCR pipelines and enabling true semantic understanding of historical records.

Connected Factories & Edge

Learning Objectives

- **Understand the Architectural Shift:** Grasp why the transition from centralized cloud AI to edge-native agentic systems is mandatory for Industry 4.0, focusing on deterministic latency and network partition tolerance.
- **Design Edge-Native Workflows:** Learn to architect autonomous agents capable of real-time reasoning over high-frequency Operational Technology (OT) data streams without relying on continuous cloud connectivity.
- **Implement Secure OT Integration:** Master the patterns for safely bridging IT and OT networks, utilizing Semantic Layer Agents to query Time-Series Databases (TSDBs) and execute actions via OPC UA and MQTT.
- **Evaluate Deployment Trade-offs:** Analyze the balance between local inference constraints, federated learning synchronization, and cloud-based digital twin simulations to optimize factory performance.

The evolution of Industry 4.0 has successfully digitized the modern shop floor, deploying millions of Industrial Internet of Things (IIoT) sensors that generate petabytes of telemetry daily. However, the cognitive architecture processing this data remains fundamentally reactive. Traditional Supervisory Control and Data Acquisition (SCADA) systems and basic rule-based engines excel at triggering alarms when a temperature threshold is breached, but they possess zero contextual awareness. When a complex, multi-variate anomaly occurs—such as a subtle vibration increase in a spindle motor correlated with a specific batch of raw materials—human operators are still required to sift through disparate dashboards, query data historians, and manually deduce the root cause. This manual bottleneck negates the speed advantages of automated manufacturing. To achieve true autonomous operations, we must transition from passive data collection to active, context-aware reasoning. This requires deploying Agentic AI directly to the factory floor—a paradigm known as *Edge-Native Agentic Architecture*.

In the context of manufacturing, Operational Technology (OT) refers to the hardware and software that detects or causes a change through the direct monitoring and/or control of physical devices, processes, and events. Historically, OT networks have been strictly segregated from Information Technology (IT) networks, governed by the ISA-95 Purdue Enterprise Reference Architecture. The Purdue Model enforces a rigid, hierarchical flow of data from physical sensors (Level 0) up to enterprise logistics (Level 4). Edge-Native Agents disrupt this hierarchy by introducing lateral, intelligent orchestration at the edge (Levels 2 and 3). Instead of streaming raw, high-frequency time-series data to a centralized cloud data lake for batch processing, *Semantic Layer Agents* reside locally. These agents are equipped with specialized tool-calling capabilities to interface directly with local Time-Series Databases (TSDBs) like OSIsoft PI or InfluxDB. They translate natural language objectives or high-level system goals into precise SQL or Flux queries, analyze the returned telemetry, and formulate execution plans without the data ever leaving the facility.

The necessity for edge deployment is driven by the unforgiving physics and security postures of the factory floor. Cloud-dependent AI architectures fail in manufacturing for three primary reasons—deterministic latency, bandwidth costs, and network partition tolerance—each of which Edge-Native Agents are architected to overcome, as summarized in Table~1.6.

Table 1.6: Why edge-native architectures are mandatory in manufacturing

Dimension	Cloud-Dependent AI	Edge-Native Agents
Deterministic latency	A round-trip API call to a cloud LLM is too slow for a welding cell at millisecond precision	Local inference enables millisecond-precision decisions on-site
Bandwidth cost	Streaming 10kHz vibration data from hundreds of machines is economically prohibitive	Telemetry analyzed locally; data never leaves the facility
Network partition tolerance	Air-gapped or intermittent links mean a network outage can halt production	Agents operate autonomously without continuous cloud connectivity

Therefore, modern industrial agents utilize lightweight, highly quantized reasoning models—such as Gemini 3.5 Flash-Lite or specialized local SLMs (Small Language Models)—deployed on industrial PCs or edge TPUs. These models provide the cognitive engine for agents to operate autonomously, ensuring continuous, secure, and low-latency decision-making.

To bridge the gap between high-level reasoning and low-level industrial protocols, developers must construct robust tool bindings. In an edge agentic framework, tools are not merely API wrappers; they are protocol translators. For example, an agent tasked with optimizing energy consumption must interface with Programmable Logic Controllers (PLCs) via protocols like OPC UA (Open Platform Communications Unified Architecture) or MQTT. The following code snippet demonstrates how an edge agent, built using a hypothetical extension of the Google Agent Development Kit (ADK) for edge environments, is configured to bind a local reasoning model to an OPC UA telemetry tool.

Listing 1.44: Configuring an Edge Agent with OPC UA Tool Bindings for Local Telemetry Access

```python
from google.adk.edge import EdgeAgent, LocalModelConfig
from google.adk.tools.industrial import OpcUaClientTool

# 1. Configure the local, quantized reasoning model (e.g., running on an Edge
    TPU)
model_config = LocalModelConfig(
    model_path="/models/gemini-3.5-flash-lite-q8.tflite",
    context_window=8192,
    temperature=0.1 # Low temperature for deterministic industrial reasoning
)

# 2. Initialize the OPC UA Tool to interface with the factory's local server
opc_tool = OpcUaClientTool(
    endpoint_url="opc.tcp://192.168.1.100:4840",
    namespace_uri="urn:factory:production_line_1",
    secure_channel=True,
    certificate_path="/certs/edge_agent.der"
)

# 3. Instantiate the Edge Agent with specific industrial instructions
historian_agent = EdgeAgent(
```

```
    name="Semantic_Historian_Agent",
    model=model_config,
    tools=[opc_tool],
    system_prompt=(
        "You are an autonomous industrial agent monitoring Production Line 1. "
        "Your goal is to detect anomalies in spindle vibration and temperature. "
        "Use the OpcUaClientTool to query real-time tags. If an anomaly is detected, "
        "correlate it with the last 10 minutes of historical data and generate a root-cause hypothesis."
    )
)

# 4. Execute a continuous monitoring loop (Sense -> Plan -> Act)
historian_agent.start_monitoring_loop(interval_ms=500)
```

While individual edge agents provide localized autonomy, a modern connected factory requires a coordinated mesh of these entities. This orchestration introduces complex challenges in lifecycle management, security, and continuous improvement. How do we update the reasoning models of hundreds of distributed agents without halting production? How do we ensure that an agent's proposed action—such as altering a machine's feed rate—is safe before it is executed? This requires a hybrid architecture where the edge handles real-time execution, while the cloud handles heavy-compute tasks like federated learning aggregation and digital twin simulation.

In this chapter, we will dissect the architectural patterns required to build, deploy, and govern this distributed intelligence. We will move beyond theoretical concepts and implement the specific agentic roles that comprise a fully autonomous factory ecosystem.

What This Chapter Covers:

- **Edge Model Fleet Manager:** Patterns for deploying, versioning, and monitoring quantized LLMs across distributed industrial PCs and edge nodes.
- **Data Historian Semantic Layer Agent:** Techniques for translating natural language queries into complex time-series database queries (SQL/Flux) to contextualize raw telemetry.
- **Cyber-OT Threat Watch (ICS):** Designing specialized security agents that monitor industrial network traffic and PLC behavior for anomalous, potentially malicious activities.
- **Federated Learning Coordinator:** Architecting systems that allow edge agents to learn from local anomalies and securely share model weight updates with the global fleet without exposing raw proprietary data.
- **Downtime Postmortem Analyst:** Implementing agents that automatically synthesize logs, operator notes, and telemetry to generate actionable root-cause analysis reports following an unplanned outage.

- **Factory Twin Scenario Simulator:** Bridging edge agents with cloud-based digital twins to safely simulate and validate proposed agentic actions before they are executed on physical hardware.

1.79 Edge Model Fleet Manager

Learning Objectives

- **Master Fleet Orchestration:** Understand the architectural necessity of a centralized control plane to manage, version, and deploy thousands of distributed edge agents across heterogeneous industrial hardware.
- **Implement Hardware-Aware Deployments:** Learn how to dynamically route specific quantized model variants (e.g., INT4, AWQ, GGUF) to edge nodes based on their specific compute capabilities (NPU, TPU, GPU).
- **Design Safe Update Mechanisms:** Architect Over-The-Air (OTA) update pipelines utilizing Shadow Deployments, Canary releases, and LoRA adapter broadcasting to update reasoning engines without halting production.
- **Establish Edge Observability:** Develop strategies for aggregating local telemetry—such as Time To First Token (TTFT), KV cache eviction rates, and semantic drift—without overwhelming constrained OT network bandwidth.

1.79.1 The Imperative for Edge Fleet Management

The transition from centralized cloud AI to edge-native agentic systems solves the critical challenges of deterministic latency, data sovereignty, and network partition tolerance. However, solving the deployment problem introduces a massive operational challenge: the "Day 2" management of distributed intelligence. In a modern Industry 4.0 facility, it is not uncommon to have hundreds, or even thousands, of Industrial PCs (IPCs), smart cameras, and robotic controllers distributed across the shop floor. If each of these endpoints hosts a local Large Language Model (LLM) or Small Language Model (SLM) acting as an autonomous agent, the factory has effectively created a highly fragmented, distributed computing cluster. Without a robust management paradigm, this cluster will rapidly degrade into an unmaintainable state of configuration drift, stale model weights, and silent failures.

Consider the lifecycle of an industrial reasoning model. A model deployed to monitor a CNC milling machine is initially fine-tuned on historical telemetry to detect tool wear. Over time, the physical dynamics of the machine change—new tooling materials are introduced, ambient factory temperatures fluctuate, and mechanical components age. The model's predictive accuracy and reasoning capabilities will inevitably experience semantic drift. To maintain operational excellence, the model must be continuously retrained or fine-tuned in the cloud (or via federated learning) and subsequently redeployed to the edge. If a factory relies on manual updates—such as technicians flashing USB drives or executing ad-hoc SSH scripts—the update cycle becomes a logistical nightmare. Furthermore, manual updates introduce unacceptable risks of human error, potentially deploying an incompatible model architecture to a constrained edge device, resulting in Out-Of-Memory (OOM) crashes and subsequent production downtime.

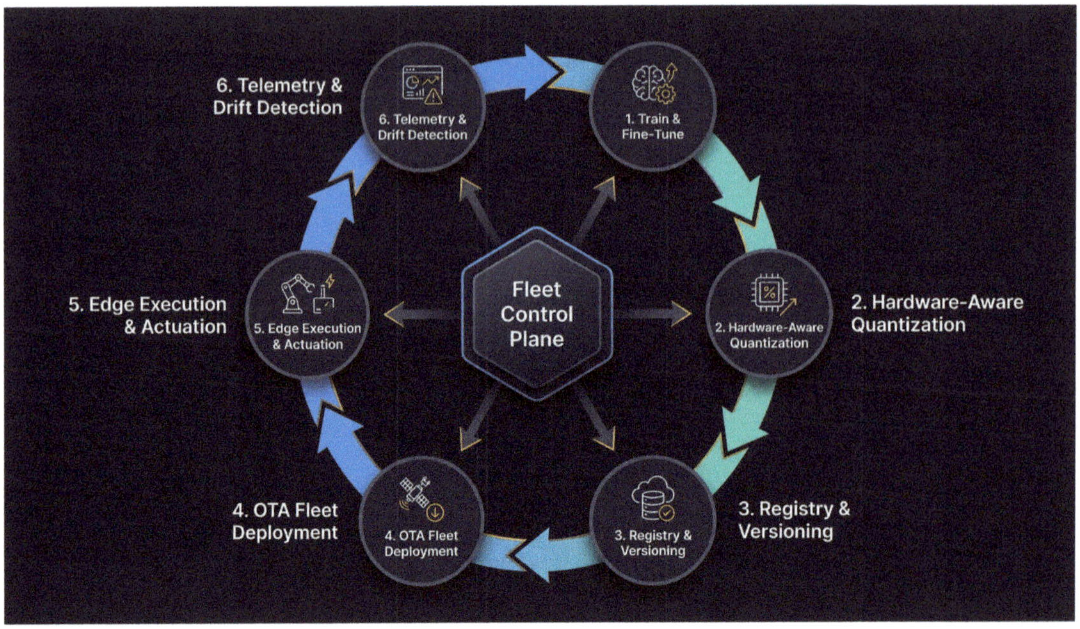

Figure 1.46: The continuous lifecycle of edge model management, orchestrated by a centralized Fleet Control Plane.

This operational reality necessitates the *Edge Model Fleet Manager*—a centralized control plane designed specifically for the lifecycle management of distributed agentic models. The Fleet Manager treats AI models not as static software binaries, but as dynamic, stateful assets that require continuous orchestration. It bridges the gap between Information Technology (IT) DevOps practices and Operational Technology (OT) constraints. In the IT world, updating a microservice is handled seamlessly by Kubernetes. In the OT world, updating an agentic model requires navigating the Purdue Enterprise Reference Architecture, traversing firewalls between Level 3 (Site Operations) and Level 2 (Area Control), and respecting the strict bandwidth limitations of industrial networks.

The Fleet Manager must address several unique constraints inherent to industrial edge environments. First is the issue of heterogeneous hardware. Unlike a cloud data center with standardized GPU racks, a factory floor is a patchwork of compute architectures. An inspection station might utilize an NVIDIA Jetson Orin for high-throughput vision processing, while a legacy PLC gateway might only have a low-power Google Coral Edge TPU or an Intel CPU with OpenVINO support. A single, monolithic model weight file cannot be deployed across this fleet. The Fleet Manager must be hardware-aware, capable of dynamically routing the correctly quantized variant of a model (e.g., an INT4 AWQ model for the GPU, and a fully integer-quantized TFLite model for the TPU) to the appropriate endpoint.

Second is the constraint of network bandwidth and connectivity. Industrial networks are optimized for low-latency, small-payload telemetry (like MQTT or OPC UA), not for transferring multi-gigabyte LLM weight files. Attempting to push a 4GB model update to 500 IPCs simultaneously over a factory's internal network will cause catastrophic congestion, potentially delaying critical safety signals. The Fleet Manager must employ intelligent distribution strategies, such as peer-to-peer (P2P) weight sharing within local subnets, chunked background downloads, and, most importantly, the broadcasting of Low-Rank Adaptation (LoRA) adapters rather than full base models. By transmitting only a 50MB LoRA adapter that modifies the behavior of a pre-existing base model, the Fleet Manager reduces network overhead by orders of magnitude.

Finally, the Fleet Manager must guarantee transactional safety. In software engineering, a failed deployment results in a 500 Internal Server Error. In manufacturing, a failed deployment can result in a robotic arm colliding with a fixture. The Fleet Manager must support atomic updates, ensuring that an edge agent only switches to a new model once the weights are fully downloaded, checksum-verified, and successfully loaded into VRAM. If the new model fails to initialize or exhibits high latency during a warm-up phase, the Fleet Manager must automatically trigger an instant rollback to the previous known-good state, ensuring zero interruption to the physical process.

1.79.2 Architecture of the Edge Model Fleet Manager

To achieve robust orchestration across thousands of nodes, the Edge Model Fleet Manager is architected around a strict separation of concerns between the *Control Plane* and the *Data Plane*. The Control Plane resides centrally—either in a secure on-premises data center (Purdue Level 3) or in a trusted cloud environment—and is responsible for policy definition, model registry, and global observability. The Data Plane consists of the actual edge nodes (Purdue Level 2 or Level 1) executing the models and interacting with the physical machinery.

At the heart of the Control Plane is the **Model Registry**. Modern fleet managers leverage Open Container Initiative (OCI) compliant registries to store model weights, tokenizers, and agent configuration files as standard container artifacts. This allows organizations to utilize existing enterprise infrastructure (like Google Artifact Registry or Harbor) to manage AI assets with the same rigorous access controls and vulnerability scanning applied to traditional software. When a data science team finalizes a new model version, it is pushed to the registry tagged with specific metadata indicating its architecture, quantization method, and target hardware profile.

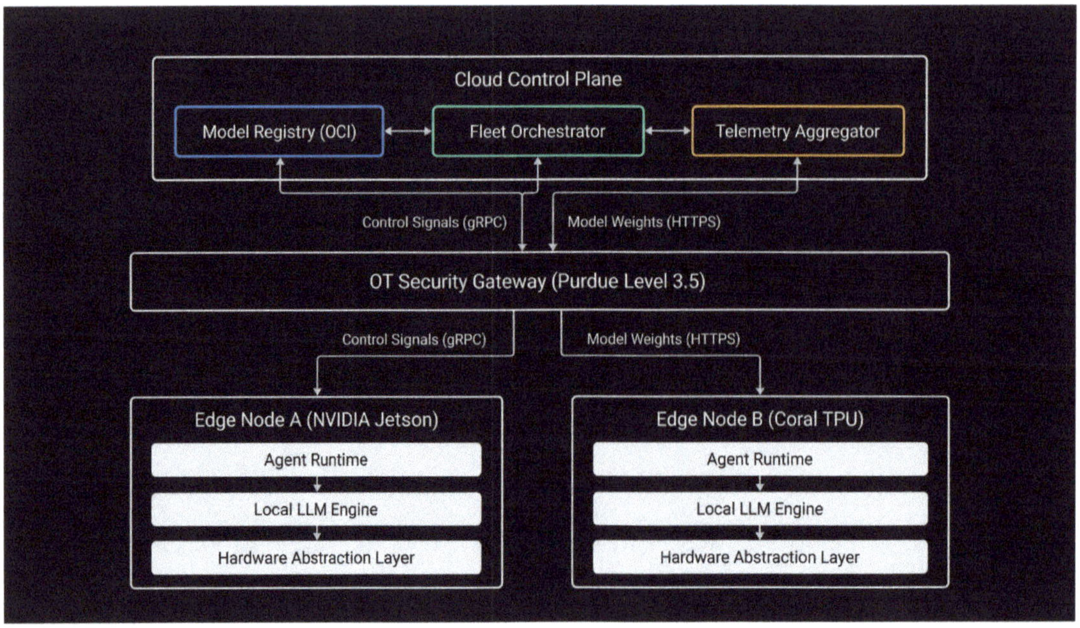

Figure 1.47: Architectural separation of the Fleet Control Plane and the Edge Data Plane, traversing OT security boundaries.

Adjacent to the registry is the **Fleet Orchestrator**. This component acts as the declarative brain of the system. Administrators define desired state manifests—for example, "All IPCs tagged with `role:welding_inspection` and `hardware:nvidia_orin` must run `gemini-3.5-flash-lite-int4:v2.1`". The Orchestrator continuously monitors the actual state of the fleet against this desired state. It communicates with the edge nodes via a lightweight, bidirectional control protocol, typically gRPC or MQTT over mutual TLS (mTLS). This control channel is strictly separated from the high-bandwidth data channel used for downloading model weights.

On the Data Plane, each IPC runs a lightweight daemon known as the **Edge Agent Runtime**. This runtime is the local representative of the Fleet Manager. It is responsible for securely authenticating with the Control Plane, reporting local hardware capabilities, and executing deployment commands. When the Orchestrator signals that a new model is required, the Edge Agent Runtime initiates a secure HTTPS pull from the Model Registry. Crucially, the runtime manages the local hardware abstraction layer. It interfaces with inference engines like `llama.cpp`, `vLLM`, or TensorRT-LLM, handling the complex orchestration of loading weights into VRAM, managing the KV cache, and exposing a local API endpoint (often OpenAI-compatible) that the actual agentic workflow scripts consume.

Capability	Manual Edge Deployment	Fleet-Managed Edge
Update Mechanism	⚠ USB / SSH Scripts	✓ Automated OTA (Over-The-Air)
Rollback Capability	⚠ Manual Intervention	✓ Instant Automated Rollback
Hardware Targeting	⚠ Homogeneous Only	✓ Heterogeneous (Auto-Quantization)
Observability	⚠ Log Scraping	✓ Centralized Telemetry Dashboard
Scale Limit	⚠ Tens of Nodes	✓ Thousands of Nodes

Figure 1.48: Capability comparison demonstrating why manual edge deployments fail to scale in enterprise manufacturing environments.

The Edge Agent Runtime is designed for extreme resilience. If the connection to the Control Plane is severed—a common occurrence in electrically noisy factory environments—the runtime continues to operate autonomously using the last known good configuration. It caches telemetry locally and synchronizes with the Control Plane once connectivity is restored. This architecture ensures that the agentic Sense-Plan-Act loop, which interacts directly with PLCs via OPC UA, remains completely decoupled from the latency and availability of the cloud.

The following code snippet illustrates how a deployment manifest is defined using a hypothetical Python SDK for the Google Agent Development Kit (ADK) Fleet Manager. This declarative approach allows MLOps engineers to manage edge models using standard Infrastructure-as-Code (IaC) practices.

Listing 1.45: Defining a Declarative Fleet Deployment Manifest for Edge Agents

```
from google.adk.fleet import FleetManager, DeploymentManifest, NodeSelector
from google.adk.fleet.models import ModelArtifact, QuantizationProfile

# Initialize connection to the Fleet Control Plane
fleet = FleetManager(
    endpoint="mTLS://fleet-control.factory.internal:8443",
    cert_path="/etc/certs/admin.pem"
)

# Define the model artifact and its hardware-specific quantization profiles
welding_model = ModelArtifact(
    registry_uri="oci://registry.factory.internal/models/weld-inspector",
    version="v2.4.0",
    profiles=[
        QuantizationProfile(hardware="nvidia_orin", format="awq", precision="
```

```
    int4"),
        QuantizationProfile(hardware="coral_tpu", format="tflite", precision="
    int8")
        ]
)

# Create a deployment manifest targeting specific nodes
manifest = DeploymentManifest(
    name="welding-cell-update-q3",
    model=welding_model,
    selector=NodeSelector(
        tags={"zone": "assembly_line_1", "role": "vision_agent"},
        min_vram_gb=4.0
    ),
    strategy="rolling_update",
    max_unavailable="10%", # Ensure 90% of nodes remain active during update
    rollback_on_failure=True
)

# Apply the manifest to the fleet
response = fleet.apply(manifest)
print(f"Deployment initiated. Tracking ID: {response.deployment_id}")
```

This declarative model shifts the burden of execution from the engineer to the system. The Fleet Orchestrator calculates the delta between the current state and the manifest, identifies the target nodes, selects the correct quantization profile for each node's specific hardware, and orchestrates the rolling update, ensuring that production capacity is never compromised.

1.79.3 Hardware-Aware Quantization and Deployment Strategies

Deploying models to the edge is fundamentally an exercise in resource constraint management. A standard 7-billion parameter LLM in FP16 precision requires approximately 14GB of VRAM just to load the weights, not accounting for the KV cache required during inference. Most industrial edge devices possess between 4GB and 8GB of shared memory. Therefore, the Fleet Manager must integrate deeply with hardware-aware quantization pipelines.

Quantization reduces the precision of the model's weights and activations (e.g., from 16-bit floats to 4-bit integers), drastically reducing memory footprint and increasing memory bandwidth utilization—the primary bottleneck in LLM inference. However, quantization is not a one-size-fits-all process. An Activation-Aware Weight Quantization (AWQ) model optimized for an NVIDIA GPU's Tensor Cores will not execute efficiently on an Intel CPU, which might prefer a GGUF format optimized for AVX-512 instructions. The Fleet Manager solves this by maintaining a matrix of hardware profiles. When a new base model is registered, the Control Plane automatically triggers a CI/CD pipeline that compiles the model into multiple target-specific formats. When an edge node requests an update, the Fleet Manager dynamically serves the artifact that matches the node's hardware signature.

Beyond quantization, the strategy used to deploy these models is critical for maintaining factory uptime. The Fleet Manager employs advanced deployment patterns adapted from cloud-native engineering, specifically tailored for physical actuation environments. The most powerful of these is the **Shadow Deployment** pattern.

In a Shadow Deployment, the Edge Agent Runtime downloads the new model version but does not immediately route live actuation commands to it. Instead, the runtime duplicates the incoming OT telemetry stream. The primary stream continues to feed the existing, trusted model, which maintains control over the PLC. The duplicated stream is fed to the new "shadow" model. The shadow model processes the telemetry and generates a proposed execution plan, but its output is routed to a local Telemetry Comparator rather than the physical machinery.

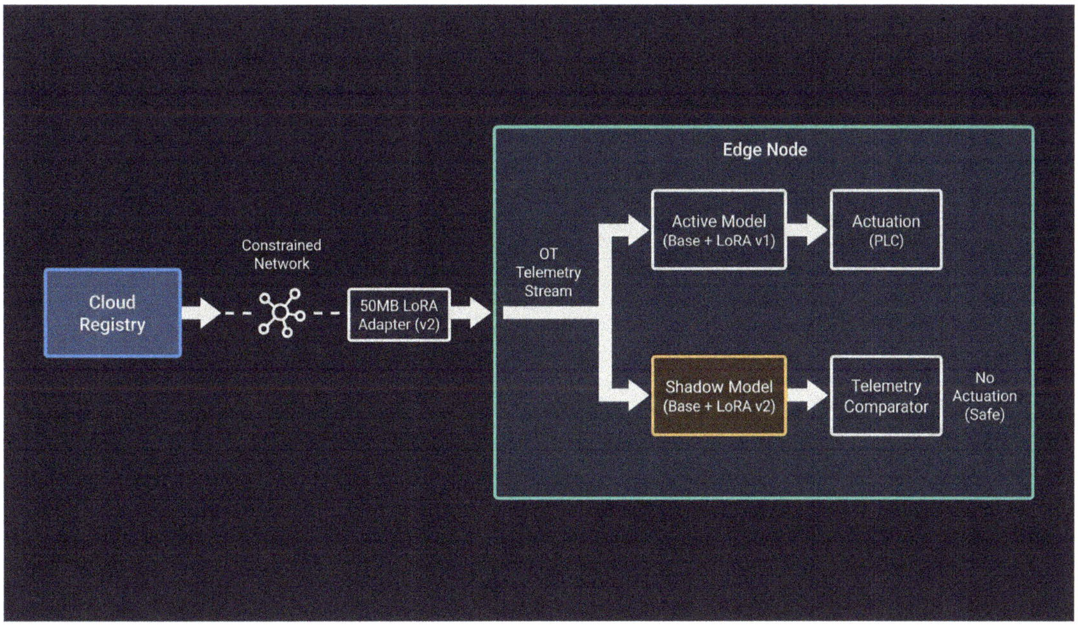

Figure 1.49: Shadow Deployment architecture utilizing LoRA adapters to safely validate new reasoning models against live OT data without risking physical actuation.

This architecture allows MLOps teams to evaluate the new model's performance against real-world, high-frequency data without risking physical damage or production halts. The Telemetry Comparator measures the divergence between the active model's decisions and the shadow model's decisions. It also monitors system-level metrics: Does the shadow model consume too much VRAM? Does its Time To First Token (TTFT) exceed the 50-millisecond deterministic threshold required by the robotic cell? If the shadow model passes all automated validation gates over a specified burn-in period (e.g., 24 hours), the Fleet Manager executes an atomic swap, promoting the shadow model to active status and deprecating the old model.

To further optimize this process, modern Fleet Managers heavily utilize **LoRA (Low-Rank Adaptation) Broadcasting**. Instead of replacing the entire multi-gigabyte base model during an update, the factory standardizes on a highly capable, quantized base model (e.g., Gemini 3.5 Flash-Lite) that is flashed onto the IPCs during initial provisioning. Subsequent updates—tailored for specific machines or newly discovered anomalies—are trained as LoRA adapters. These adapters represent only the delta in weights and are typically tens of megabytes in size. The Fleet Manager broadcasts these tiny adapters over the OT network. The Edge Agent Runtime dynamically loads the adapter into VRAM alongside the base model, instantly altering the agent's behavior with near-zero network impact. This enables rapid, continuous iteration of agentic intelligence at the edge.

1.79.4 Observability, Telemetry, and Drift Detection at the Edge

Deploying models to the edge is only half the battle; ensuring they continue to operate correctly over time is equally challenging. In a cloud environment, every prompt, response, and intermediate reasoning step can be logged to a centralized data lake for analysis. In an edge environment, this is impossible. Streaming the full context windows of hundreds of autonomous agents back to the cloud would instantly saturate the factory's network backbone. Therefore, the Fleet Manager must implement an intelligent, edge-local observability paradigm.

Edge observability relies on the principle of *Federated Telemetry Aggregation*. Instead of streaming raw logs, the Edge Agent Runtime continuously calculates statistical summaries of model performance and agent behavior. These metrics are categorized into two distinct domains: System Telemetry and Semantic Telemetry.

System Telemetry focuses on the computational health of the inference engine. The runtime monitors metrics such as Time To First Token (TTFT), Tokens Per Second (TPS), VRAM utilization, and KV cache eviction rates. In an industrial setting, these metrics are not merely performance indicators; they are safety constraints. If an agent controlling a high-speed sorting conveyor experiences a sudden spike in TTFT due to KV cache saturation, it may miss a critical decision window, leading to a physical jam. The Edge Agent Runtime is configured with hard latency thresholds. If TPS drops below the required threshold, the runtime can automatically shed non-critical context (e.g., summarizing older memory) or, in extreme cases, trigger a graceful degradation protocol, handing control back to a deterministic, rule-based PLC fallback.

Semantic Telemetry is far more complex. It attempts to measure the quality and safety of the agent's reasoning. Because the agent is operating autonomously, we must detect when it begins to hallucinate or when its understanding of the physical environment drifts from reality. To achieve this without cloud connectivity, the Edge Agent Runtime employs a *Local Critic Model*. This is a highly quantized, ultra-small model (often under 1 billion parameters) whose sole purpose is to evaluate the outputs of the primary reasoning agent.

The Critic Model operates on a strict rubric. For example, if the primary agent generates a JSON payload intended for the OPC UA tool to change a machine's temperature setpoint, the Critic Model intercepts the payload before execution. It verifies that the proposed temperature is within the hardcoded safety bounds of the machine's physical specifications. Furthermore, the runtime calculates confidence scores based on the primary model's token log-probabilities. If the agent proposes a high-impact physical action but exhibits low confidence (high perplexity) in its generated tokens, the runtime flags the action as anomalous.

Listing 1.46: Edge-Local Telemetry Aggregation and Semantic Drift Detection

```
import time
import numpy as np
from google.adk.edge.observability import TelemetryAggregator, CriticEvaluator

class EdgeAgentMonitor:
    def __init__(self):
        self.aggregator = TelemetryAggregator(flush_interval_minutes=15)
        # Load a tiny, fast critic model for local safety checks
        self.critic = CriticEvaluator(model_path="/models/safety-critic-int8.
    tflite")
```

```python
    def monitor_inference(self, prompt, generated_plan, token_logprobs,
    latency_ms):
        # 1. Record System Telemetry
        self.aggregator.record_metric("latency_ms", latency_ms)

        # Calculate confidence based on log probabilities
        avg_confidence = np.exp(np.mean(token_logprobs))
        self.aggregator.record_metric("model_confidence", avg_confidence)

        # 2. Perform Semantic Safety Check via Local Critic
        safety_score = self.critic.evaluate(
            context=prompt,
            proposed_action=generated_plan,
            rubric="Ensure temperature setpoint does not exceed 250C."
        )

        if safety_score < 0.95 or avg_confidence < 0.80:
            self.aggregator.increment_counter("safety_interventions")
            # Block actuation and trigger fallback
            return False

        return True

    def sync_with_fleet_manager(self, fleet_client):
        # Only send statistical summaries to the cloud, not raw prompts
        summary_stats = self.aggregator.get_summaries()
        fleet_client.push_telemetry(summary_stats)
```

Periodically (e.g., every 15 minutes), the Edge Agent Runtime flushes these aggregated statistical summaries—average latency, confidence distributions, and intervention counts—over the control channel to the Fleet Manager's Telemetry Aggregator in the cloud. This approach requires mere kilobytes of bandwidth.

At the Control Plane, MLOps engineers monitor these fleet-wide dashboards. If a specific cluster of edge nodes (e.g., all agents on Assembly Line 3) begins showing a gradual decline in average confidence scores or an increase in safety interventions, it is a definitive signal of semantic drift. The physical environment has likely changed in a way the model does not understand. The Fleet Manager can then automatically trigger a data collection campaign, instructing those specific edge nodes to temporarily stream a sample of their raw prompts and OT telemetry back to the cloud. This targeted data is used to fine-tune a new LoRA adapter, which is subsequently broadcasted back to the fleet, closing the continuous improvement loop of the autonomous factory.

Key Takeaways

- **Centralized Control, Decentralized Execution:** A Fleet Manager is mandatory for scaling edge AI, providing a cloud-based control plane to orchestrate models while ensuring edge nodes retain autonomous execution capabilities during network partitions.

- **Hardware-Awareness is Non-Negotiable:** Industrial environments feature highly heterogeneous compute. The Fleet Manager must dynamically route specifically quantized models (INT4, INT8, AWQ) to match the exact NPU/TPU/GPU profile of the target edge device.
- **LoRA Broadcasting Minimizes Network Strain:** Instead of pushing multi-gigabyte base models over constrained OT networks, utilize LoRA adapters to transmit only the behavioral deltas, reducing update payloads to tens of megabytes.
- **Shadow Deployments Ensure Safety:** Validate new reasoning models by running them in parallel with active models on live OT telemetry streams, comparing their proposed actions without allowing them to actuate physical machinery.
- **Federated Telemetry Prevents Bandwidth Saturation:** Edge nodes must aggregate system metrics (TTFT, VRAM) and semantic metrics (confidence scores, local critic evaluations) locally, sending only statistical summaries back to the cloud to detect model drift.

Learning Objectives

- Understand the "semantic gap" between raw Operational Technology (OT) time-series data and Large Language Model reasoning capabilities.
- Architect a Knowledge Graph-backed semantic layer utilizing industrial ontologies (e.g., ISA-95) to contextualize historian tags.
- Implement a Data Historian Agent capable of translating natural language into complex, multi-step time-series queries (e.g., PI-SQL, InfluxQL).
- Design safety guardrails, query validators, and data summarization techniques to prevent context window saturation and historian database overload.
- Deploy agentic workflows for autonomous root cause analysis and temporal correlation across disparate manufacturing assets.

1.80 The Semantic Gap in Operational Technology (OT)

The lifeblood of any modern manufacturing facility flows through its data historian. Systems like OSIsoft PI (now AVEVA PI System), Ignition, and InfluxDB are engineered to ingest, compress, and store millions of data points per second from Programmable Logic Controllers (PLCs), Supervisory Control and Data Acquisition (SCADA) systems, and Industrial Internet of Things (IIoT) sensors. These historians are the undisputed source of truth for factory operations, holding decades of high-fidelity time-series data that describe every temperature fluctuation, pressure drop, and motor vibration that has ever occurred on the plant floor.

However, when we attempt to connect a reasoning engine—such as a Gemini 3.5 Pro agent—directly to a data historian, we immediately encounter a profound architectural barrier known as the *semantic gap*.

Data historians are fundamentally optimized for write-speed and storage density, not for human or machine comprehension. The data is stored as flat lists of "tags" (or points). In a typical enterprise facility, there may be upwards of 500,000 individual tags. The naming conventions for these tags are often archaic, inconsistent, and heavily abbreviated, reflecting the physical wiring diagrams or legacy DCS (Distributed Control System) configurations rather than the logical function of the equipment.

For example, a tag might be named `B14_ZIC_9021.PV`. To a veteran plant engineer, this string might intuitively mean "Building 14, Zone Isolation Controller, Valve 9021, Present Value." To an LLM, it is an opaque, meaningless string of characters. If an operator asks an agent, "Why did the primary cooling loop on Extruder Line 3 fail yesterday at 14:00?", the agent cannot answer. It has no inherent way to know which of the 500,000 tags correspond to "Extruder Line 3," which tags represent the "primary cooling loop," or what constitutes a "failure" in the context of those specific sensors.

To bridge this gap, we must introduce a **Semantic Layer**—an architectural pattern that abstracts the underlying data structures into a business-friendly, logically organized format. In the context of Agentic AI, this semantic layer is typically implemented as an Industrial Knowledge Graph (IKG). The graph acts as a translation engine, mapping the physical reality of the factory (the tags) to the logical reality of the business (the assets, lines, and processes).

By equipping an agent with tools to traverse this semantic layer, we transform it from a blind text-generator into a highly capable Data Historian Agent. This agent can dynamically discover the correct data sources, formulate precise time-series queries, retrieve the historical data, and apply advanced reasoning to diagnose complex operational issues.

1.81 Architecting the Industrial Knowledge Graph (IKG)

The foundation of the Data Historian Semantic Layer Agent is the Knowledge Graph. Unlike relational databases, which struggle with deeply nested hierarchical relationships, graph databases (such as Neo4j, Amazon Neptune, or Google Cloud Spanner Graph) are purpose-built for modeling the complex, interconnected nature of a manufacturing plant.

To ensure interoperability and maintainability, the graph should be structured according to established industrial ontologies. The most prevalent standard is **ISA-95** (IEC 62264), which defines a strict equipment hierarchy: *Enterprise* $\rightarrow$ *Site* $\rightarrow$ *Area* $\rightarrow$ *Line* $\rightarrow$ *Work Cell* $\rightarrow$ *Unit*.

In our Knowledge Graph, these hierarchical levels become **Nodes**. The relationships between them become **Edges** (e.g., `HAS_AREA`, `CONTAINS_LINE`). Crucially, the raw historian tags are also represented as nodes in the graph, connected to the physical equipment they monitor via edges like `MONITORS_TEMPERATURE` or `CONTROLS_PRESSURE`.

Consider the modeling of a simple cooling system:

- **Node (Asset):** `Extruder_3`
- **Node (Asset):** `Cooling_Pump_A`
- **Node (Tag):** `EXT3_CP_A_FLOW.PV` (The actual historian tag ID)
- **Edge:** `Extruder_3 [-FED_BY->] Cooling_Pump_A`
- **Edge:** `Cooling_Pump_A [-HAS_SENSOR->] EXT3_CP_A_FLOW.PV`

When the agent receives a prompt regarding "Extruder 3's cooling flow," it does not guess the tag name. Instead, it executes a graph traversal query (e.g., using Cypher) to find the node

Extruder_3, traverse the FED_BY edge to find the cooling equipment, and traverse the HAS_-SENSOR edge to retrieve the exact string EXT3_CP_A_FLOW.PV. Only then does it construct the SQL or Flux query to hit the actual time-series database.

This two-stage retrieval process—Semantic Resolution followed by Temporal Retrieval—is the defining design pattern of the Data Historian Agent. It completely decouples the LLM's reasoning process from the brittle, hardcoded tag structures of the legacy OT environment. If a sensor is replaced and its tag name changes from EXT3_CP_A_FLOW.PV to EXT3_CP_A_FLOW_-V2.PV, the MLOps team simply updates the single edge in the Knowledge Graph. The agent's prompts, tools, and reasoning logic remain entirely unchanged.

1.82 Implementing the Historian Query Agent

Building this system requires orchestrating multiple tools within an agentic framework like LangGraph, CrewAI, or the Google Agent Development Kit (ADK). The agent must be equipped with specific tools for graph traversal and time-series extraction, along with strict system prompts that dictate the sequence of operations.

A naive implementation might simply give the LLM a tool that executes raw SQL against the historian. This is a catastrophic anti-pattern. Time-series databases are highly sensitive to poorly constructed queries. An LLM hallucinating a query like SELECT * FROM tags WHERE time > '2010-01-01' without proper downsampling or aggregation will attempt to pull billions of rows into memory, instantly crashing the historian server and taking down factory visibility.

Therefore, the tools provided to the agent must be heavily constrained, parameterized, and wrapped in safety validators. The following Python code demonstrates how to implement a robust, safe toolset for a Data Historian Agent using a parameterized approach.

Listing 1.47: Implementing Safe Semantic and Time-Series Tools for the Historian Agent

```python
from typing import List, Dict, Any
from pydantic import BaseModel, Field
from datetime import datetime, timedelta
import pandas as pd

# 1. Define the input schema for the Semantic Resolution Tool
class AssetResolutionInput(BaseModel):
    asset_name: str = Field(..., description="The natural language name of the
     equipment (e.g., 'Extruder 3')")
    measurement_type: str = Field(..., description="The type of measurement
    needed (e.g., 'temperature', 'pressure', 'vibration')")

# 2. Define the input schema for the Time-Series Retrieval Tool
class TimeSeriesQueryInput(BaseModel):
    tag_ids: List[str] = Field(..., description="List of exact historian tag
    IDs to query.")
    start_time: str = Field(..., description="ISO 8601 start time.")
    end_time: str = Field(..., description="ISO 8601 end time.")
    aggregation_window: str = Field(default="15m", description="Downsampling
    window (e.g., '1m', '15m', '1h'). Required to prevent data overload.")
    aggregation_function: str = Field(default="mean", description="Function to
     apply (mean, max, min, last).")
```

```python

class HistorianTools:
    def __init__(self, graph_client, historian_client):
        self.graph = graph_client
        self.historian = historian_client

    def resolve_asset_to_tags(self, input_data: AssetResolutionInput) -> Dict[str, Any]:
        """Tool 1: Resolves natural language asset names to exact historian tag IDs."""
        # Construct a safe, parameterized Cypher query
        cypher_query = """
        MATCH (a:Asset {name: $asset_name})-[:HAS_SENSOR]->(s:Sensor)
        WHERE s.type = $measurement_type
        RETURN s.tag_id AS tag_id, s.description AS description
        """
        results = self.graph.execute(
            cypher_query,
            parameters={"asset_name": input_data.asset_name, "measurement_type": input_data.measurement_type}
        )

        if not results:
            return {"error": f"No tags found for {input_data.measurement_type} on {input_data.asset_name}."}

        return {"resolved_tags": [r["tag_id"] for r in results], "context": results}

    def fetch_historian_data(self, input_data: TimeSeriesQueryInput) -> str:
        """Tool 2: Safely fetches and summarizes time-series data."""
        # SAFETY GUARDRAIL 1: Enforce maximum time range
        start = datetime.fromisoformat(input_data.start_time.replace('Z', '+00:00'))
        end = datetime.fromisoformat(input_data.end_time.replace('Z', '+00:00'))
        if (end - start) > timedelta(days=30):
            return "Error: Query exceeds maximum allowed time range of 30 days. Please narrow your search."

        # SAFETY GUARDRAIL 2: Enforce aggregation
        if input_data.aggregation_window not in ["1m", "5m", "15m", "1h", "1d"]:
            input_data.aggregation_window = "15m" # Force safe default

        # Construct the parameterized query (e.g., InfluxQL or Flux)
        # The client handles the actual execution safely
        try:
            df = self.historian.query_aggregated(
                tags=input_data.tag_ids,
                start=start,
                end=end,
                window=input_data.aggregation_window,
                agg_func=input_data.aggregation_function
```

```
            )

            # SAFETY GUARDRAIL 3: Prevent Context Window Saturation
            # Never return raw dataframes to the LLM if they are too large.
            if len(df) > 500:
                # Return statistical summary instead of raw rows
                summary = df.describe().to_markdown()
                return f"Data too large for raw output. Statistical summary:\n{summary}"

            return df.to_markdown()

        except Exception as e:
            return f"Historian query failed: {str(e)}"
```

Notice the critical design decisions in this implementation. The agent is never allowed to write raw SQL. Instead, it populates a strictly typed Pydantic schema (`TimeSeriesQueryInput`). The underlying Python code enforces safety guardrails: it limits the maximum time range to 30 days, it forces data aggregation (downsampling) to prevent the retrieval of millions of raw data points, and it monitors the size of the resulting Pandas DataFrame. If the DataFrame exceeds 500 rows, the tool automatically intercepts the payload and returns a statistical summary (`df.describe()`) rather than flooding the LLM's context window with raw numbers, which would degrade reasoning performance and spike token costs.

1.83 Temporal Reasoning and Root Cause Analysis

With the semantic layer and safe retrieval tools in place, the Data Historian Agent unlocks a capability that has historically required hours of manual engineering effort: autonomous Root Cause Analysis (RCA) across disparate machine centers.

In a traditional setup, if a quality defect is detected at the end of a production line, an engineer must manually open a historian client (like PI Vision), guess which upstream machines might have caused the issue, manually search for their tag names, drag those tags onto a trend chart, and visually scan for anomalies that occurred hours prior.

An agentic system automates this entire investigative loop. Because the Knowledge Graph contains the physical topology of the plant (e.g., the `FEEDS_INTO` edges), the agent can perform temporal correlation by walking backward through the production process.

Consider the prompt: *"We had a high-porosity defect on Line 1 at 14:30. Find the root cause."*

The agent's reasoning loop (Sense → Plan → Act → Reflect) executes as follows: 1. **Plan:** The agent determines it needs to find the equipment upstream of the defect point, identify critical process parameters (CPPs) for that equipment, and query the historian for anomalies leading up to 14:30. 2. **Act (Tool Call 1):** The agent queries the Knowledge Graph for the topology of Line 1. The graph returns that `Mixer_A` feeds into `Extruder_1`, which feeds into the defect inspection station. 3. **Act (Tool Call 2):** The agent queries the graph for the critical tags on `Mixer_A` and `Extruder_1`. It retrieves tags for mixer RPM, extruder barrel temperature, and cooling water pressure. 4. **Act (Tool Call 3):** The agent queries the Historian Tool for these specific tags between 13:30 and 14:30 (accounting for the physical residence time of the material moving through the line). 5. **Reflect:** The agent analyzes the returned markdown table. It notes that extruder temperatures were normal, but `Mixer_A` RPM dropped by 15% at 14:12. 6. **Act (Tool Call 4):** The agent queries the graph to find what powers `Mixer_A`. It discovers `VFD_Drive_7`. 7. **Act (Tool Call 5):** It queries the historian for `VFD_Drive_7` fault codes at 14:12. It retrieves a value corresponding to an "Over-temperature Warning." 8. **Final Output:** The agent synthesizes the findings into a natural language report: *"The high-porosity defect at 14:30 was likely caused by a 15% drop in Mixer A's RPM at 14:12. This RPM drop correlates exactly with an over-temperature warning on VFD Drive 7. Recommend inspecting the cooling fan on VFD Drive 7."*

This multi-step reasoning process is impossible with standard RAG (Retrieval-Augmented Generation) over text documents. It requires the deterministic structure of the Knowledge Graph combined with the temporal depth of the Data Historian, orchestrated by an LLM capable of complex tool chaining.

1.84 Production Considerations and Pitfalls

Deploying a Data Historian Semantic Layer Agent into a live manufacturing environment requires strict adherence to enterprise governance and performance optimization patterns.

Read-Only Enforcement: The most critical safety rule is that the agent's connection to the historian must be strictly read-only. Historians are regulated systems (often subject to FDA 21 CFR Part 11 in pharmaceuticals or NERC CIP in energy). An agent must never be granted permissions to write, alter, or delete historical data. All tool connections should use dedicated service accounts with database-level read-only roles. If the agent needs to take action based on its findings (e.g., creating a work order in SAP PM), that action must be routed through a separate, explicitly governed tool with Human-in-the-Loop (HITL) approval gates.

Context Window Management and Token Costs: Time-series data is incredibly token-dense. A single hour of data from 10 sensors sampled at 1-second intervals yields 36,000 rows. Passing this raw data into an LLM context window will result in massive latency, exorbitant API costs, and severe degradation in the model's ability to reason (the "lost in the middle" phenomenon). As demonstrated in the code example, the agent must rely on the historian's native compute engine to perform aggregations (min, max, mean, standard deviation) *before* the data is returned to the LLM. The LLM should only process statistical summaries or highly downsampled trend lines.

Handling Graph Drift: The Semantic Layer is only as useful as its accuracy. In a dynamic factory, sensors are replaced, lines are reconfigured, and PLCs are reprogrammed. If the physical reality drifts from the Knowledge Graph, the agent will hallucinate, querying the wrong tags or failing to find assets. To mitigate this, MLOps teams must implement automated graph synchronization pipelines. These pipelines continuously ingest the configuration files from the DCS or SCADA systems (e.g., parsing OPC UA node spaces) and automatically update the edges in the Neo4j or Spanner Graph database, ensuring the agent's map of the territory is always perfectly aligned with reality.

Key Takeaways

- **The Semantic Gap is the Primary Blocker:** LLMs cannot reason over raw, cryptically named historian tags. A Semantic Layer is mandatory to map physical data points to logical business assets.
- **Use Industrial Ontologies:** Structure the Knowledge Graph using established standards like ISA-95 to ensure the agent understands the hierarchical relationships (Site → Line → Equipment → Sensor).
- **Implement Two-Stage Retrieval:** Design agent tools to first query the graph to resolve natural language to tag IDs, and then query the historian to retrieve the temporal data.
- **Enforce Strict Query Guardrails:** Never allow the LLM to write raw SQL. Use parameterized schemas that enforce maximum time ranges and mandatory data downsampling to protect the historian from overload.
- **Summarize Before Reasoning:** Time-series data will overwhelm context windows. Push aggregations down to the historian database and only return statistical summaries or downsampled dataframes to the agent.

Learning Objectives

- Understand the unique constraints of Industrial Control Systems (ICS) and Operational Technology (OT) security, including the Purdue Enterprise Reference Architecture and the "do no harm" principle.
- Architect a multi-agent Security Operations Center (SOC) capable of correlating raw network telemetry with physical process data from the Data Historian.
- Implement an Agentic Threat Watch system that maps anomalous behaviors to the MITRE ATT&CK for ICS framework to reduce false positives.
- Design Human-in-the-Loop (HITL) response gateways that ensure autonomous agents cannot inadvertently cause physical damage or process downtime.

In the previous chapter, we explored how the Data Historian Semantic Layer Agent bridges the gap between cryptic, low-level sensor tags and high-level business logic, enabling autonomous root cause analysis for manufacturing defects. We now turn our attention to the most critical, high-stakes application of agentic systems in the industrial sector: securing the physical processes themselves.

The convergence of Information Technology (IT) and Operational Technology (OT) has shattered the traditional "air gap" that once protected factories, power plants, and water treatment facilities. As edge devices, Industrial Internet of Things (IIoT) sensors, and cloud-connected Programmable Logic Controllers (PLCs) proliferate, the attack surface of the modern connected factory has expanded exponentially. Defending these environments requires a paradigm shift from reactive, signature-based IT security to proactive, context-aware Agentic Threat Watch systems.

1.85 The IT/OT Convergence and the ICS Threat Landscape

To understand why agentic AI is necessary for OT security, we must first understand the structural differences between IT and OT environments. Industrial networks are traditionally modeled using the Purdue Enterprise Reference Architecture (PERA), which segments systems into hierarchical levels based on their function and criticality.

At the top are Levels 4 and 5 (Enterprise IT), where standard corporate networks, email servers, and ERP systems reside. Level 3 encompasses Site Operations, including the Data Historians and Manufacturing Execution Systems (MES) discussed previously. Level 2 contains the Supervisory Control systems, such as Human-Machine Interfaces (HMIs) and SCADA servers. Level 1 houses the basic control devices—the PLCs and Remote Terminal Units (RTUs) that execute the control logic. Finally, Level 0 represents the physical process itself: the valves, motors, pumps, and robotic arms.

Historically, a strict firewall—the IT/OT Demilitarized Zone (DMZ)—separated Level 3 from Level 4. However, the drive for Industry 4.0 analytics, predictive maintenance, and remote monitoring has necessitated the flow of data from Level 1 directly to the cloud, effectively bypassing traditional Purdue segmentation. This convergence exposes fragile, legacy OT devices to modern cyber threats.

The fundamental challenge in securing ICS environments is the "semantic gap." Traditional IT Security Information and Event Management (SIEM) tools, such as Splunk or Microsoft Sentinel, are designed to monitor IP addresses, ports, and Windows event logs. If an IT SIEM detects a connection over TCP port 502 (Modbus), it simply logs a network flow. It has no understanding of the payload.

However, in an OT environment, a Modbus `Write Multiple Registers` command sent from an unauthorized IP address is not just a network anomaly; it is a physical event. It could be a command instructing a centrifuge to spin beyond its physical tolerances (as seen in the Stuxnet attack), or a command disabling a Safety Instrumented System (SIS) prior to a destructive payload (as seen in the Triton/Trisis attack).

When IT SIEMs are deployed in OT environments without physical context, they generate thousands of false positives. An operator logging into an HMI to adjust a setpoint looks identical to an attacker manipulating that same setpoint. The result is alert fatigue; security analysts ignore the dashboard, and critical threats slip through.

This is where Agentic AI transforms the landscape. An Agentic Threat Watch system does not merely look at network packets; it reasons over the relationship between the network traffic, the asset inventory, and the physical state of the process. By integrating the Data Historian Semantic Layer (built in the previous chapter) with network intrusion detection systems (like Zeek or Suricata), the agent can autonomously ask: *"I see a Modbus write command changing the pressure setpoint on Valve 4. Is this normal behavior for this time of day? Who issued the command? And most importantly, what is the current physical pressure reading in the historian?"*

1.86 Architecting the Cyber-OT Agentic SOC

Building an Agentic Threat Watch system requires a multi-agent architecture. A single monolithic LLM prompt cannot handle the sheer volume of network telemetry, parse proprietary industrial protocols, query the historian, and map findings to the MITRE ATT&CK for ICS framework simultaneously. Instead, we must decompose the problem into specialized agentic roles orchestrated by a state graph.

The architecture consists of three primary agents:

1. **The Network Analysis Agent:** This agent is equipped with tools to query network telemetry databases (e.g., Elasticsearch indices containing Zeek logs). It specializes in parsing industrial protocols (Modbus, DNP3, CIP, S7 Comm) and identifying anomalous communication patterns, such as a Level 4 IT workstation attempting to communicate directly with a Level 1 PLC.

2. **The Process Context Agent:** This agent leverages the Knowledge Graph and Data Historian tools developed in the previous chapter. Its role is to take an IP address or MAC address flagged by the Network Agent, resolve it to a physical asset (e.g., "Extruder 3 PLC"), and query the historian to determine if the physical process is exhibiting anomalous behavior (e.g., unexpected temperature spikes).

3. **The Triage Commander Agent:** This is the orchestrator. It receives the findings from the Network and Process agents, maps the combined behavior to specific Tactics, Techniques, and Procedures (TTPs) in the MITRE ATT&CK for ICS matrix, scores the severity of the threat, and drafts a containment playbook.

To implement this architecture robustly, we utilize a stateful framework like LangGraph. The state of the investigation must be maintained as the agents iteratively gather context. The following Python code demonstrates how to define the state schema and orchestrate the multi-agent workflow.

Listing 1.48: Defining the Multi-Agent OT SOC Workflow using LangGraph

```python
from typing import TypedDict, List, Dict, Any, Annotated
from langgraph.graph import StateGraph, END
from langchain_core.messages import BaseMessage, HumanMessage, AIMessage
import operator

# 1. Define the Investigation State
class OTInvestigationState(TypedDict):
    alert_id: str
    raw_network_event: Dict[str, Any]
    network_analysis: str
    identified_assets: List[str]
    process_context: str
    mitre_ics_mapping: List[str]
    severity_score: int
    containment_playbook: str
    messages: Annotated[List[BaseMessage], operator.add]

# 2. Define the Agent Nodes
def network_analysis_node(state: OTInvestigationState):
    """Analyzes the raw Zeek/Suricata alert and extracts protocol specifics."""
    event = state["raw_network_event"]
    # In production, this invokes an LLM with tools to query PCAP/Zeek
```

```
    databases
    prompt = f"Analyze this OT network event: {event}. Identify the protocol, source/dest IPs, and any anomalous function codes (e.g., unauthorized Modbus writes)."

    # Simulated LLM response for demonstration
    analysis = "Detected Modbus TCP Write Multiple Registers (Function Code 16) from unauthorized IT subnet (10.0.5.x) to PLC at 192.168.1.50."
    assets = ["192.168.1.50"]

    return {"network_analysis": analysis, "identified_assets": assets}

def process_context_node(state: OTInvestigationState):
    """Queries the Historian and Knowledge Graph for physical context."""
    assets = state["identified_assets"]
    # In production, this invokes the Data Historian Agent from the previous chapter
    prompt = f"For assets {assets}, query the Knowledge Graph to identify the physical equipment. Then query the Historian for the last 15 minutes of data to detect physical anomalies."

    # Simulated LLM response
    context = "IP 192.168.1.50 resolves to 'Cooling Pump PLC'. Historian data shows pump RPM dropped to 0 immediately following the Modbus write command. Physical process is degrading."

    return {"process_context": context}

def triage_commander_node(state: OTInvestigationState):
    """Synthesizes findings, maps to MITRE ICS, and drafts a playbook."""
    net_analysis = state["network_analysis"]
    proc_context = state["process_context"]

    # In production, this invokes an LLM with a strict system prompt for MITRE mapping
    mapping = ["T0836: Modify Parameter", "T0889: Modify Control Logic"]
    severity = 95 # Critical
    playbook = "1. Isolate IT subnet 10.0.5.x at the Level 3 firewall. 2. Dispatch operator to Cooling Pump PLC to switch to manual override. 3. Do NOT reboot PLC until memory is dumped for forensics."

    return {
        "mitre_ics_mapping": mapping,
        "severity_score": severity,
        "containment_playbook": playbook
    }

# 3. Build the Graph
workflow = StateGraph(OTInvestigationState)

workflow.add_node("NetworkAgent", network_analysis_node)
workflow.add_node("ProcessAgent", process_context_node)
workflow.add_node("TriageCommander", triage_commander_node)
```

```
workflow.set_entry_point("NetworkAgent")
workflow.add_edge("NetworkAgent", "ProcessAgent")
workflow.add_edge("ProcessAgent", "TriageCommander")
workflow.add_edge("TriageCommander", END)

# Compile the agentic application
ot_soc_app = workflow.compile()
```

This state graph ensures a deterministic flow of information. The Network Agent cannot hallucinate physical impacts because it does not have access to the historian tools. The Process Agent cannot misinterpret network protocols because it relies on the Network Agent's output. The Triage Commander acts as the synthesizer, ensuring the final output is actionable and grounded in both cyber and physical reality.

1.87 Autonomous Alert Triage and Physical Contextualization

The true power of the Agentic Threat Watch system lies in its ability to drastically reduce false positives through physical contextualization. In a traditional SOC, an analyst receiving an alert for an "Unauthorized Modbus Write" has to make a judgment call based purely on network data. If they block the traffic and it turns out to be a legitimate engineering update, they cause an unplanned factory outage—a career-limiting move in the OT world. If they ignore it and it is a real attack, the physical process could be destroyed.

By integrating the Process Context Agent, the system performs the exact same verification steps a senior OT engineer would perform, but in milliseconds. Consider a scenario where an engineering workstation (Level 2) sends a firmware update command to a Safety Instrumented System (SIS) controller (Level 1).

A traditional SIEM flags this as a high-severity alert because firmware updates are rare and highly privileged. The Agentic Threat Watch system intercepts this alert and triggers the Process Context Agent. The agent queries the Knowledge Graph and determines that the target IP is indeed the SIS controller for the primary boiler. It then queries the Data Historian for the boiler's current state.

If the historian reports that the boiler is currently operating at 800 degrees Celsius and 150 PSI, the agent recognizes a critical anomaly: *Legitimate firmware updates are never pushed to an active, running safety controller.* The agent immediately escalates the alert to a Severity 100, maps it to MITRE ICS Tactic TA0111 (Inhibit Response Function), and flags it as a likely Triton-style attack.

Conversely, if the historian reports that the boiler's temperature is ambient, the pressure is zero, and the "Maintenance Mode" tag is set to TRUE, the agent contextualizes the network event as a legitimate, scheduled maintenance activity. It suppresses the alert, logging it for audit purposes but preventing it from waking up the on-call engineer at 3:00 AM. This contextual suppression is the key to solving alert fatigue in OT environments.

1.88 Safe Response Planning and Human-in-the-Loop (HITL) Execution

While autonomous detection and triage represent a massive leap forward, autonomous *response* in OT environments remains highly controversial. In IT security, if an agent detects ransomware on a laptop, the standard automated response is to isolate the machine from the network by altering switch port configurations or updating endpoint detection and response (EDR) policies.

In OT, automated isolation can be more dangerous than the cyberattack itself. If an agent autonomously isolates a PLC that is actively controlling an exothermic chemical reaction, the loss of visibility and control could lead to a catastrophic physical explosion. The "do no harm" principle is paramount. Therefore, Agentic Threat Watch systems must be designed with strict Human-in-the-Loop (HITL) gateways for any active response.

The role of the Triage Commander Agent is not to execute the response, but to generate a highly specific, context-aware *Containment Playbook* for human review. This playbook must include not only the cyber actions (e.g., "Block IP 10.0.5.45 at Firewall FW-02") but also the physical prerequisites (e.g., "Verify Line 2 is in a safe shutdown state before executing firewall block").

To implement this safely, we utilize the Agent-to-Agent (A2A) or Agent Payments Protocol (AP2) concepts discussed in Part II of this book, adapting them for authorization rather than commerce. When the Triage Commander generates a playbook, it submits a "Request for Action" to a HITL Gateway. This gateway pauses the agent's execution thread and routes a rich, interactive notification to the human SOC operator (via Microsoft Teams, Slack, or a dedicated SOC dashboard).

The notification contains the full context: the raw network alert, the historian data proving physical impact, the MITRE mapping, and the proposed playbook. The human operator can click "Approve," "Deny," or "Modify." If approved, the gateway returns a cryptographic token to the agent, unlocking its execution tools (e.g., an API connection to the Palo Alto firewall) to carry out the block. If denied, the agent logs the human feedback into its semantic memory, learning from the correction to improve future triage accuracy.

This architecture ensures that the speed and analytical power of the LLM are fully leveraged for investigation, while the ultimate authority and liability for physical safety remain firmly in human hands. As the system matures and trust is established, organizations can gradually transition specific, low-risk playbooks (e.g., isolating a known-bad external IP address at the perimeter) to fully autonomous execution, while keeping critical internal OT actions strictly behind the HITL gateway.

Key Takeaways

- **The Semantic Gap is the Enemy:** Traditional IT SIEMs fail in OT because they lack physical context. Agentic systems bridge this gap by correlating network packets with physical historian data.
- **Multi-Agent Orchestration is Required:** Decompose the SOC into specialized agents (Network Analysis, Process Context, Triage Commander) using a state graph to ensure deterministic, reliable investigations.

- **Contextual Suppression Cures Alert Fatigue:** By verifying the physical state of the machinery, agents can autonomously suppress false positives (e.g., legitimate maintenance activities) that would otherwise overwhelm human operators.
- **Safety First, Autonomy Second:** Never allow an agent to autonomously isolate OT equipment without human approval. The "do no harm" principle dictates that agents generate playbooks, but humans authorize execution.
- **Implement Strict HITL Gateways:** Use cryptographic authorization tokens and interactive approval workflows to ensure human operators maintain ultimate control over the physical process.

Learning Objectives

- Understand the architectural shift from centralized model training to Agentic Federated Learning in Operational Technology (OT) environments.
- Design a multi-agent orchestration system capable of dynamic edge cohort selection, straggler mitigation, and asynchronous weight aggregation.
- Implement secure aggregation protocols governed by LLM-driven anomaly detection to prevent model poisoning and Byzantine failures.
- Deploy a Federated Learning Coordinator using LangGraph, integrating edge-native communication protocols (MQTT/OPC UA) with centralized global model state management.

1.89 The Imperative for Agentic Federated Learning in Connected Factories

In the preceding chapter, we explored the architecture of the Cyber-OT Agentic Security Operations Center (SOC), establishing how intelligent agents can monitor, contextualize, and secure the physical processes of a connected factory. However, security is only one dimension of the Industry 4.0 paradigm. The ultimate promise of the connected factory is predictive optimization: utilizing machine learning to foresee equipment failures, optimize energy consumption, and dynamically adjust manufacturing parameters to maximize yield. Achieving this requires training sophisticated deep learning models on the telemetry generated by the factory floor.

Historically, the dominant paradigm for training these models has been centralized. Data from Programmable Logic Controllers (PLCs), Industrial PCs (IPCs), and Internet of Things (IoT) sensors is streamed to a centralized data lake—often hosted in a public cloud environment like Google Cloud or AWS—where massive compute clusters train the models. Once trained, the inference models are pushed back down to the edge. While conceptually simple, this centralized approach fundamentally breaks down in modern, high-throughput manufacturing environments due to three insurmountable barriers: data gravity, intellectual property (IP) protection, and network latency.

First, consider the sheer volume of data generated by modern industrial equipment. A single Computer Numerical Control (CNC) milling machine equipped with high-frequency vibration sensors on its spindle might sample data at 50 kHz across three axes. Using 64-bit floating-point precision, this single sensor generates approximately 1.2 megabytes of data per second.

A factory floor with one hundred such machines generates over 10 terabytes of raw vibration data per day. Attempting to backhaul this volume of data over a factory's local area network (LAN), through the corporate wide area network (WAN), and into a cloud environment incurs exorbitant bandwidth costs and rapidly saturates available network infrastructure. This phenomenon, known as "data gravity," dictates that compute must move to the data, rather than moving the data to the compute.

Second, raw manufacturing telemetry is highly sensitive intellectual property. The exact vibration signatures, thermal profiles, and torque curves of a manufacturing process can be reverse-engineered to reveal proprietary manufacturing techniques, material compositions, and production volumes. Many manufacturers, particularly in aerospace, defense, and high-tech electronics, operate under strict data residency and compliance regulations (such as ITAR in the United States) that explicitly forbid raw production data from leaving the physical premises of the factory. Centralized cloud training is often a non-starter from a legal and compliance perspective.

Third, the latency introduced by centralized training cycles prevents real-time adaptation. If a new batch of raw material exhibits slightly different metallurgical properties, the edge models must adapt their predictive maintenance thresholds immediately. Waiting for a daily or weekly batch upload to the cloud, followed by a retraining cycle and a subsequent model deployment, leaves the factory vulnerable to unplanned downtime during the interim period.

Federated Learning (FL) emerged as the cryptographic and architectural solution to these challenges. In a standard FL paradigm, the raw data never leaves the edge device. Instead, a global model is distributed to the edge nodes (e.g., the IPCs attached to the CNC machines). Each edge node trains the model locally using its own private, high-frequency data. After a set number of local epochs, the edge nodes transmit only the updated model weights (or gradients) back to a central server. The central server aggregates these weights—typically using an algorithm like Federated Averaging (FedAvg)—to create a new, improved global model, which is then redistributed to the edge. The IP remains secure, bandwidth consumption is reduced by orders of magnitude, and the models benefit from the collective intelligence of the entire fleet.

However, standard Federated Learning is highly rigid and assumes a level of network and data homogeneity that simply does not exist in a real-world factory. Standard FL algorithms struggle profoundly with Non-Independent and Identically Distributed (Non-IID) data. In a factory, Machine A might be milling titanium while Machine B is milling aluminum; their local data distributions are vastly different. If the central server blindly averages their weights, the resulting global model may perform poorly on both. Furthermore, standard FL assumes reliable compute and network availability. If Machine C is currently executing a highly complex, CPU-intensive control loop, it may not have the compute cycles available to participate in the training round, causing the entire synchronous aggregation process to stall—a phenomenon known as the "straggler problem."

This is where the transition from static algorithms to *Agentic* Federated Learning becomes critical. An Agentic Federated Learning Coordinator is an LLM-driven orchestration system that dynamically manages the FL lifecycle. Instead of executing a rigid, hardcoded loop, the Coordinator agent continuously senses the state of the factory floor. It evaluates the data drift on individual machines, assesses the current CPU and network load of the edge IPCs, and dynamically selects a cohort of edge nodes for the next training round. It reasons about the statistical distribution of the incoming weights, identifying and quarantining anomalous updates that could indicate a compromised node or a catastrophic sensor failure. By applying multi-agent reasoning to the orchestration of distributed machine learning, we transform a fragile mathematical protocol into a robust, adaptive, and enterprise-grade system capable of operating in the chaotic environment of the connected factory.

1.90 Architecture of the Agentic Federated Learning Coordinator

Designing an Agentic Federated Learning system requires a departure from traditional monolithic machine learning pipelines. The architecture must be distributed, stateful, and capable of asynchronous operation. To achieve this, we utilize a multi-agent topology orchestrated by a state graph framework, such as LangGraph, deployed on the Gemini Enterprise Agent Platform. This architecture separates the concerns of local execution, global orchestration, and security evaluation into distinct, specialized agentic roles.

The architecture comprises three primary components: the Edge Training Agents, the Global Aggregator Agent, and the Security & Evaluation Agent.

1. The Edge Training Agents (Local Actors): These agents reside directly on the industrial edge—typically deployed as containerized workloads on IPCs or specialized edge AI gateways (e.g., NVIDIA Jetson devices) located adjacent to the manufacturing equipment. Unlike passive FL clients that simply wait for a command from the server, Edge Training Agents are proactive. They continuously monitor the local inference performance of the current model. By calculating metrics such as prediction error, confidence degradation, or statistical data drift (using techniques like the Kolmogorov-Smirnov test on incoming sensor streams), the Edge Agent determines when local retraining is necessary.

When drift exceeds a predefined threshold, the Edge Agent does not immediately begin training. Instead, it sends a "Request to Train" message to the central Coordinator via a lightweight industrial protocol like MQTT. This request includes metadata about the edge node's current state: available CPU/GPU compute, memory availability, local dataset size, and a statistical summary of the data distribution (e.g., class balance or feature histograms). This metadata is crucial for the central Coordinator to make informed orchestration decisions.

2. The Global Aggregator Agent (The Coordinator): Residing in the central factory data center or a secure virtual private cloud (VPC), the Global Aggregator Agent is the orchestrator of the entire system. It maintains the global state of the federated learning process, including the current version of the global model weights, the registry of all active edge nodes, and the historical performance metrics of the fleet.

The Aggregator Agent operates a continuous reasoning loop. It ingests the "Request to Train" messages from the Edge Agents and evaluates them against the global objective. If enough edge nodes are experiencing drift, or if a scheduled training interval has elapsed, the Aggregator initiates a new Federated Learning round. Its most critical function is *Dynamic Cohort Selection*—choosing exactly which edge nodes should participate in the current round based on their compute availability and data distributions, ensuring a balanced and efficient training cycle. Once the cohort is selected, the Aggregator dispatches the current global model weights and training hyperparameters to the selected Edge Agents.

3. The Security & Evaluation Agent (The Gatekeeper): As the Edge Agents complete their local training epochs, they transmit their updated model weights (or weight deltas) back to the central server. Before these weights are merged into the global model, they must pass through the Security & Evaluation Agent. This agent acts as a cryptographic and statistical gatekeeper. It analyzes the incoming weight vectors for signs of model poisoning, catastrophic forgetting, or hardware-induced corruption.

Instead of relying solely on rigid mathematical thresholds, this agent utilizes an LLM to reason about the context of the updates. For example, if an edge node submits a weight update with a massive gradient norm, a standard FL system might blindly accept it (ruining the global model) or blindly reject it. The Security Agent, however, can query the factory's Data Historian (as discussed in Chapter 285) to check the physical state of that specific machine. If the historian confirms that the machine recently underwent a major mechanical overhaul, the massive shift in the model weights is contextually justified, and the agent may allow the update. If no physical event justifies the shift, the update is quarantined as a potential Byzantine failure or cyberattack.

To implement this orchestration robustly, we define the global state and the agentic workflow using LangGraph. The state must capture the complexities of distributed training, including the tracking of asynchronous updates. The following Python code demonstrates the foundational state schema and the node definitions for the Global Aggregator and Security Agents.

Listing 1.49: Defining the Agentic Federated Learning Coordinator State in LangGraph

```
from typing import TypedDict, List, Dict, Any, Annotated
from langgraph.graph import StateGraph, END
import operator

# 1. Define the Federated Learning Global State
class FLCoordinatorState(TypedDict):
    round_number: int
    global_model_version: str
    available_edge_nodes: Dict[str, Dict[str, Any]] # Node ID -> Telemetry
    selected_cohort: List[str]
    received_updates: Dict[str, Any] # Node ID -> Weight Deltas/Metrics
    quarantined_nodes: List[str]
    aggregation_strategy: str # e.g., 'FedAvg', 'FedProx'
    round_status: str
    audit_log: Annotated[List[str], operator.add]

# 2. Define the Agent Nodes
def cohort_selection_node(state: FLCoordinatorState):
    """Analyzes edge telemetry and selects the optimal training cohort."""
    nodes = state["available_edge_nodes"]

    # In production, an LLM or heuristic evaluates compute/data distribution
    # Here we simulate selecting nodes with CPU load < 80% and high data drift
    selected = []
    for node_id, telemetry in nodes.items():
        if telemetry.get("cpu_load", 100) < 80 and telemetry.get("
    data_drift_score", 0) > 0.5:
            selected.append(node_id)

    if len(selected) < 3: # Minimum cohort size constraint
        return {"round_status": "ABORTED_INSUFFICIENT_NODES", "audit_log": ["
```

```
    Aborted: Not enough healthy nodes."]}

    return {
        "selected_cohort": selected,
        "round_status": "TRAINING_DISPATCHED",
        "audit_log": [f"Round {state['round_number']}: Dispatched to {len(
    selected)} nodes."]
    }

def security_evaluation_node(state: FLCoordinatorState):
    """Evaluates incoming weight updates for anomalies or poisoning."""
    updates = state["received_updates"]
    quarantined = state.get("quarantined_nodes", [])
    valid_updates = {}

    # Simulate LLM/Statistical evaluation of weight deltas
    for node_id, update_data in updates.items():
        l2_norm = update_data.get("gradient_l2_norm", 0)

        # If the gradient norm is suspiciously high, quarantine the update
        if l2_norm > 15.0:
            quarantined.append(node_id)
            state["audit_log"].append(f"SECURITY ALERT: Node {node_id}
    quarantined due to anomalous L2 norm ({l2_norm}).")
        else:
            valid_updates[node_id] = update_data

    return {
        "received_updates": valid_updates, # Pass only valid updates forward
        "quarantined_nodes": quarantined,
        "round_status": "EVALUATION_COMPLETE"
    }

def aggregation_node(state: FLCoordinatorState):
    """Aggregates the validated weights to create the new global model."""
    valid_updates = state["received_updates"]

    if len(valid_updates) == 0:
        return {"round_status": "FAILED_ALL_QUARANTINED"}

    # In production, this triggers PyTorch/TensorFlow to perform FedAvg/
    FedProx
    new_version = f"v{state['round_number'] + 1}.0"

    return {
        "global_model_version": new_version,
        "round_number": state["round_number"] + 1,
        "round_status": "AGGREGATION_COMPLETE",
        "audit_log": [f"Successfully aggregated {len(valid_updates)} updates
    into {new_version}."]
    }

# 3. Build the Orchestration Graph
workflow = StateGraph(FLCoordinatorState)
```

```
workflow.add_node("CohortSelector", cohort_selection_node)
workflow.add_node("SecurityEvaluator", security_evaluation_node)
workflow.add_node("Aggregator", aggregation_node)

workflow.set_entry_point("CohortSelector")
# Routing logic based on status
workflow.add_edge("CohortSelector", "SecurityEvaluator")
workflow.add_edge("SecurityEvaluator", "Aggregator")
workflow.add_edge("Aggregator", END)

fl_app = workflow.compile()
```

This state graph ensures that the orchestration of the federated learning process is deterministic, auditable, and resilient. The state object (`FLCoordinatorState`) acts as the single source of truth, allowing the system to recover gracefully if the central server experiences a transient failure during a training round.

1.91 Dynamic Cohort Selection and Straggler Mitigation

One of the most profound challenges in operationalizing Federated Learning in industrial environments is the "straggler problem." In a standard, synchronous FL round, the central server dispatches the model to N edge nodes and waits for all N nodes to return their updated weights before performing the aggregation. If a single node experiences a network dropout, or if its CPU is suddenly monopolized by a critical real-time control task (e.g., an emergency braking sequence on a robotic arm), that node becomes a straggler. The entire global training process halts, waiting indefinitely for the delayed update. In a factory with hundreds of nodes, the probability of at least one straggler occurring in any given round approaches 100%.

Traditional FL frameworks attempt to solve this with simple timeouts: if a node doesn't respond within T seconds, drop it and aggregate the rest. However, this naive approach introduces severe statistical bias. Often, the nodes that are dropped are the ones processing the most complex, anomalous, or high-volume data—exactly the data the global model needs to learn from the most. Dropping them systematically leads to a degraded model that performs poorly on edge cases.

The Agentic FL Coordinator solves this through *Predictive Dynamic Cohort Selection* and *Asynchronous Aggregation*.

Instead of randomly selecting nodes or relying on static timeouts, the Cohort Selection Agent utilizes an LLM to reason over the real-time telemetry of the entire fleet before a round begins. The agent evaluates a multi-dimensional optimization problem: it must select a cohort that maximizes data diversity (handling Non-IID distributions) while minimizing the probability of stragglers (based on compute and network forecasts).

For example, the agent receives telemetry indicating that IPC-A has a 90% CPU load and a highly unstable Wi-Fi connection, but it contains a rare dataset of tool-wear anomalies. IPC-B has a 10% CPU load and a stable ethernet connection, but its data is entirely nominal (no anomalies). A naive timeout system would select IPC-A, wait for it, and eventually drop it, losing the rare data. The Agentic Coordinator, however, reasons differently. It recognizes IPC-

A's compute constraints and explicitly excludes it from the synchronous cohort. Instead, it issues a specialized, asynchronous training task to IPC-A with a relaxed deadline, allowing IPC-A to compute the gradients slowly in the background without blocking the main global round.

Furthermore, the agent actively balances the statistical distribution of the cohort. If the factory has 50 machines producing Part X and 10 machines producing Part Y, a random selection will heavily bias the model toward Part X. The Cohort Selection Agent analyzes the metadata summaries provided by the edge nodes and intentionally over-samples the minority class (Part Y machines) to ensure the aggregated global model generalizes well across all production lines.

When dealing with highly heterogeneous data, the agent may also dynamically switch the underlying mathematical aggregation strategy. While Federated Averaging (FedAvg) is the default, it performs poorly when local data distributions diverge significantly. If the agent detects high divergence in the data summaries, it can instruct the Aggregator Node to utilize Federated Proximal (FedProx) instead. FedProx introduces a proximal term to the local objective function, which restricts the local updates from drifting too far from the global model, ensuring stable convergence even in highly Non-IID industrial environments. The ability of the LLM to dynamically select the mathematical optimization strategy based on real-time factory conditions is a hallmark of true agentic autonomy.

1.92 Secure Aggregation and Defense Against Model Poisoning

As Federated Learning pushes model training to the absolute edge of the network, it inherently expands the attack surface of the AI system. In a centralized architecture, the training data and compute infrastructure are secured within a highly controlled data center. In an FL architecture, the training occurs on IPCs distributed across the factory floor—devices that are physically accessible and often connected to legacy, less-secure OT networks.

This distributed nature introduces the risk of *Byzantine failures* and *Model Poisoning*. A Byzantine failure occurs when an edge node behaves arbitrarily or maliciously, sending corrupted weight updates back to the central server. This could be due to a hardware fault (e.g., a failing RAM module corrupting the gradient tensors) or a deliberate cyberattack. In a model poisoning attack, an adversary compromises an edge node and intentionally manipulates its local training data (data poisoning) or directly alters the computed gradients (model poisoning) before transmission. The goal is to introduce a backdoor into the global model—for example, teaching the predictive maintenance model to systematically ignore the vibration signatures of a specific, impending catastrophic failure, thereby causing physical damage to the factory.

Because the central server in an FL system never sees the raw data, it cannot easily verify the legitimacy of the weight updates. If the Aggregator Agent blindly averages a poisoned update with the legitimate updates, the global model becomes compromised. Defending against this requires the Security & Evaluation Agent to implement robust, Byzantine-fault-tolerant (BFT) aggregation protocols.

The Security Agent employs a multi-layered defense strategy. The first layer is statistical anomaly detection. Before aggregation, the agent calculates the pairwise distances (e.g., cosine similarity or Euclidean distance) between all incoming weight update vectors. In a healthy cohort, even with Non-IID data, the updates should generally point in a similar direction in the high-dimensional parameter space, as they are all optimizing toward the same global objective. If one node's update vector is orthogonal to the rest, or if its L2 norm is orders of magnitude larger, it is mathematically highly suspicious.

Instead of relying on simple averaging, the agent can dynamically deploy robust aggregation algorithms such as Krum or Trimmed Mean. The Krum algorithm, for instance, selects a single update vector that is closest to its $N-f$ neighbors (where f is the assumed number of Byzantine nodes) and uses that as the global update, effectively discarding extreme outliers.

However, sophisticated attackers can craft "stealthy" poisoning attacks that keep the gradient norms small enough to evade basic statistical filters while still introducing the backdoor. To counter this, the Security Agent utilizes its LLM reasoning capabilities combined with a *Centralized Validation Set*. The central server maintains a small, highly curated, and cryptographically secured dataset of known-good and known-bad factory telemetry.

Before finalizing the aggregation, the Security Agent temporarily applies the proposed global model update and evaluates its performance against this trusted validation set. If the model's accuracy on detecting a specific, critical anomaly suddenly drops from 99% to 50% after applying the update, the agent immediately detects the catastrophic forgetting or poisoning attempt.

When a poisoned update is detected, the Agentic Coordinator does not simply drop the update and move on. It executes a security playbook. It quarantines the offending edge node, preventing it from participating in future rounds. Crucially, it interfaces with the Cyber-OT SOC (detailed in Chapter 286), sending an automated alert containing the node's IP address, the mathematical proof of the anomaly, and a request to isolate the IPC at the network level for forensic investigation. This seamless integration between the Machine Learning operations (MLOps) pipeline and the Security Operations Center (SOC) exemplifies the power of a unified, agentic enterprise architecture. The AI system is not only optimizing the factory; it is actively defending itself.

Key Takeaways

- **Overcoming Data Gravity:** Federated Learning is essential in Industry 4.0 because the sheer volume and sensitivity of high-frequency OT telemetry make centralized cloud training technically and legally unfeasible.
- **Agentic Orchestration is Required:** Standard FL fails in factories due to Non-IID data and stragglers. An Agentic Coordinator dynamically manages the training lifecycle, adapting to real-time compute and network constraints.
- **Predictive Cohort Selection:** By reasoning over edge telemetry, the Coordinator agent can assemble balanced training cohorts and delegate resource-constrained nodes to asynchronous background queues, eliminating the straggler problem.
- **Defending the Global Model:** Pushing training to the edge expands the attack surface. The system must employ a Security Agent to perform statistical anomaly detection (e.g., Krum, Trimmed Mean) on incoming weight vectors to prevent Byzantine failures and model poisoning.
- **Validation and SOC Integration:** Advanced poisoning attacks are mitigated by testing aggregated updates against a trusted central validation set. Quarantined nodes must automatically trigger alerts in the Cyber-OT SOC for physical and network investigation.

Learning Objectives

- Understand the architectural shift from reactive, human-led Root Cause Analysis (RCA) to autonomous, agentic postmortem workflows in Industry 4.0 environments.
- Design a multi-agent diagnostic architecture capable of fusing structured Operational Technology (OT) telemetry with unstructured Information Technology (IT) maintenance logs.
- Implement a LangGraph-based reasoning engine that autonomously queries industrial historians, formulates hypotheses, and validates evidence.
- Evaluate the trade-offs between context window limitations, token costs, and retrieval-augmented generation (RAG) when processing high-frequency factory data.
- Deploy human-in-the-loop (HITL) guardrails to ensure the safety and accuracy of AI-generated corrective actions in mission-critical manufacturing.

1.93 The Anatomy of Agentic Root Cause Analysis

In the high-stakes domain of modern manufacturing and Industry 4.0, unplanned downtime is the ultimate adversary. For a high-volume automotive assembly line or a continuous-process chemical plant, a single minute of unexpected stoppage can incur costs ranging from tens to hundreds of thousands of dollars. When a critical asset fails—be it a robotic welding cell, a high-speed packaging conveyor, or a multi-axis CNC machining center—the immediate priority is restoring production. However, the secondary, and arguably more critical, priority is ensuring that the specific failure mode never occurs again. This is the domain of Root Cause Analysis (RCA), a process traditionally executed through structured methodologies such as the 8D (Eight Disciplines) framework, the 5-Whys, or Ishikawa (Fishbone) diagrams.

Historically, the postmortem process has been a highly manual, labor-intensive, and cognitively demanding exercise. When a machine faults, a cross-functional team comprising reliability engineers, maintenance technicians, and process operators must convene. They manually extract alarm logs from the Supervisory Control and Data Acquisition (SCADA) system, export time-series telemetry from the industrial historian (such as OSIsoft PI or Ignition), pull maintenance histories from the Computerized Maintenance Management System (CMMS) like SAP PM or IBM Maximo, and review operator shift notes. This data gathering phase alone can consume days. Once the data is aggregated, the team relies on human intuition and experience to correlate disparate signals—for example, linking a subtle voltage drop on a Tuesday to a catastrophic servo motor failure on a Thursday. This human-led approach is inherently flawed. It is susceptible to cognitive biases (such as confirmation bias, where engineers look only for data that supports their initial hunch), it is constrained by the limited capacity of human working memory to process multi-variate time-series data, and it is fundamentally unscalable across a global fleet of thousands of connected machines.

The advent of Large Language Models (LLMs) and, more specifically, Agentic AI architectures, fundamentally transforms this paradigm. The Downtime Postmortem Analyst is not merely a text-generation tool that summarizes a provided log file; it is an autonomous, goal-seeking system designed to orchestrate the entire diagnostic workflow. When triggered by a critical fault code from the Manufacturing Execution System (MES), the agentic system wakes up, assumes the persona of a master reliability engineer, and begins its investigation. It possesses the capability to dynamically write and execute SQL queries against the CMMS, formulate API calls to the historian to extract specific time windows of sensor data, and utilize Retrieval-Augmented Generation (RAG) to consult thousands of pages of OEM (Original Equipment Manufacturer) technical manuals and past RCA reports.

To understand why an agentic approach is necessary, we must examine the concept of IT/OT convergence. Operational Technology (OT) encompasses the hardware and software that detects or causes a change through the direct monitoring and/or control of physical devices, processes, and events in the enterprise. Information Technology (IT) encompasses the systems used for data processing, storage, and enterprise resource planning. A true root cause often spans both domains. For instance, a machine might fail (OT event) because a procurement system automatically switched to a cheaper, out-of-spec raw material supplier (IT event). A human engineer might take weeks to connect these dots across siloed departments. An agentic system, equipped with cross-domain tool access and a unified semantic understanding of the enterprise, can traverse this graph of causality in minutes.

The architecture of the Downtime Postmortem Analyst relies heavily on the ReAct (Reasoning and Acting) pattern, augmented with a Plan-and-Solve meta-strategy. Upon receiving a fault notification, the agent does not immediately begin guessing. Instead, it generates an explicit investigation plan. It might reason: "Step 1: Retrieve the exact alarm sequence 5 minutes prior to the fault. Step 2: Query the CMMS to see when this specific asset was last lubricated. Step 3: If lubrication was recent, check the historian for abnormal temperature spikes in the spindle bearing. Step 4: Cross-reference findings with the OEM troubleshooting guide." This explicit planning phase is crucial for interpretability. In a highly regulated manufacturing environment (such as aerospace or pharmaceuticals), the AI cannot be a black box. The reliability engineering team must be able to audit the agent's chain of thought to trust its conclusions.

Furthermore, the agentic postmortem process is inherently asynchronous. Unlike real-time control agents (which must operate in the millisecond regime to prevent physical collisions), the postmortem agent operates in the minutes-to-hours regime. This relaxed latency constraint allows the system to utilize frontier reasoning models (such as Gemini 3.5 Pro) and execute deep, multi-step reflection loops. If the agent queries the historian and receives a flatline signal (indicating a broken sensor rather than a process anomaly), it has the time and cognitive capacity to recognize the bad data, critique its own initial hypothesis, and pivot its investigation toward an electrical failure or a severed communication cable. This capacity for self-correction and dynamic replanning elevates the system from a simple automated script to a true autonomous analyst.

1.94 Multi-Modal Data Fusion and Hypothesis Generation

The core cognitive challenge for the Downtime Postmortem Analyst is multi-modal data fusion. The agent must synthesize insights from fundamentally different data structures: high-frequency numerical time-series data (vibration, temperature, pressure), structured relational data (maintenance schedules, part numbers, vendor IDs), and unstructured text (operator shift logs, PDF manuals, past RCA narratives). Feeding this heterogeneous mix directly into the context window of an LLM is an anti-pattern that leads to severe hallucination, context dilution, and exorbitant token costs.

Consider the time-series data from an industrial historian. A single vibration sensor on a CNC spindle might sample at 10 kHz. Over a 10-minute window leading up to a crash, this generates 6 million floating-point numbers. No current LLM can or should process this raw array. Instead, the agentic system must employ Context Engineering and specialized tool invocation. The agent does not ask for the raw data; it asks a specialized Python tool to compute statistical features. The agent might invoke a tool called `analyze_vibration_trend(asset_id="CNC_04", start_time="T-10m", end_time="T-0")`. The underlying Python execution environment queries the historian, runs a Fast Fourier Transform (FFT) or calculates the Root Mean Square (RMS), kurtosis, and peak-to-peak values, and returns a semantic summary to the LLM: *"The vibration RMS increased exponentially from 2.1 mm/s to 15.4 mm/s over the 10-minute window, with a dominant frequency emerging at 120 Hz."* The LLM can now reason over this semantic insight, recognizing 120 Hz as a classic signature of mechanical unbalance or bearing wear.

Once the initial data is gathered, the agent enters the Hypothesis Generation phase. This is where the system leverages its vast pre-trained knowledge of physics, engineering principles, and industrial mechanics, grounded by the specific context of the factory. The agent utilizes a Tree-of-Thought (ToT) or Instruction Tree pattern to systematically explore potential failure modes.

For example, if the initial symptom is a "Main Drive Overcurrent Fault," the agent generates a mutually exclusive and collectively exhaustive (MECE) set of hypotheses. Hypothesis A: Mechanical binding (the motor is working too hard to turn a stuck physical component). Hypothesis B: Electrical short (insulation failure causing a current spike). Hypothesis C: Operational error (the operator commanded a feed rate that exceeded the machine's design limits).

The true power of the agentic system is that it does not stop at generating hypotheses; it actively seeks to falsify them. To test Hypothesis C, the agent queries the MES for the active part program and the commanded feed rate at the time of the crash. If the feed rate was within nominal limits, Hypothesis C is pruned. To test Hypothesis B, the agent queries the CMMS for recent megger test results (insulation resistance) on that specific motor. If the tests were nominal, the probability of Hypothesis B decreases. To test Hypothesis A, the agent queries the historian for the torque and vibration trends leading up to the overcurrent event. If it finds the exponential vibration increase mentioned earlier, it confirms Hypothesis A and drills down further: what caused the mechanical binding?

This iterative process of hypothesis generation, evidence retrieval, and hypothesis pruning mimics the scientific method. It is vastly superior to traditional rule-based expert systems. A rule-based system operates on rigid `IF-THEN` logic (e.g., `IF Overcurrent AND Vibration > 10 THEN Bearing Failure`). Rule-based systems are brittle; they fail silently when presented with novel failure modes or edge cases not explicitly programmed by a human. An agentic system, powered by a frontier reasoning model, can interpolate between known concepts and deduce novel failure mechanisms by reasoning from first principles.

Furthermore, the agent utilizes RAG to ground its hypotheses in authoritative documentation. When investigating the bearing failure, the agent queries the vector database containing the OEM manuals. It retrieves the specific maintenance schedule for that bearing type. If the manual dictates lubrication every 500 operating hours, and the agent's previous CMMS query revealed the asset had run for 800 hours since the last lubrication work order, the root cause is definitively established: a failure in the preventive maintenance execution process. The agent has successfully traced a physical OT failure (vibration/overcurrent) back to an IT/process failure (missed work order).

1.95 Autonomous Diagnostic Workflows and Tool Execution

To operationalize this cognitive process, we must translate the conceptual architecture into a deterministic, stateful workflow. Frameworks like LangGraph are ideally suited for this task, as they allow developers to define the agent's reasoning process as a cyclical graph, managing state explicitly and providing strict control over tool execution and routing logic.

In a LangGraph implementation of the Downtime Postmortem Analyst, the state object (`DowntimeState`) acts as the central repository of truth throughout the investigation. It holds the initial fault trigger, the accumulated evidence from various tools, the current list of active hypotheses, and the final drafted report. The workflow is divided into specialized nodes, each representing a distinct phase of the investigation or a specific sub-agent.

Below is a production-grade implementation of the core orchestration logic using Python and LangGraph. This example demonstrates how the agent transitions from data gathering to hypothesis generation, and crucially, how it employs an `EvidenceValidator` node to act as a critic, ensuring the generator does not hallucinate conclusions without supporting data.

Listing 1.50: LangGraph Implementation of the Downtime Postmortem Analyst

```
import operator
from typing import Annotated, List, Dict, Any, Optional
from typing_extensions import TypedDict
from langgraph.graph import StateGraph, END
from langchain_core.messages import HumanMessage, AIMessage, SystemMessage

# 1. Define the State Schema
class DowntimeState(TypedDict):
    asset_id: str
    fault_code: str
    timestamp: str
    raw_alarms: List[str]
    gathered_evidence: Annotated[List[Dict[str, Any]], operator.add]
    active_hypotheses: List[str]
    validation_feedback: str
    final_rca_report: Optional[str]
    iteration_count: int

# 2. Define the Nodes (Sub-Agents and Logic)
def data_gatherer_node(state: DowntimeState):
    """Autonomously queries Historian and CMMS based on the fault code."""
    # In production, this node uses an LLM to decide WHICH tools to call
    # Here we simulate the tool execution returning structured evidence
    evidence = []
    if "Overcurrent" in state["fault_code"]:
        # Simulated tool call: query_historian(state["asset_id"], metric="
    vibration")
        evidence.append({"source": "Historian", "metric": "vibration_rms", "
    trend": "exponential_increase", "peak": "15.4 mm/s"})
        # Simulated tool call: query_cmms(state["asset_id"], record_type="
```

```python
maintenance")
        evidence.append({"source": "CMMS", "last_lube_date": "2025-11-01", "status": "overdue_by_300_hours"})

    return {"gathered_evidence": evidence, "iteration_count": state.get("iteration_count", 0) + 1}

def hypothesis_generator_node(state: DowntimeState):
    """Formulates hypotheses based on the gathered evidence."""
    # Prompt the LLM to generate hypotheses
    evidence_str = str(state["gathered_evidence"])
    prompt = f"Asset {state['asset_id']} faulted with {state['fault_code']}. Evidence: {evidence_str}. Generate 3 plausible root causes."

    # Simulated LLM response
    hypotheses = [
        "H1: Bearing failure due to lack of lubrication.",
        "H2: Electrical short in the motor windings.",
        "H3: Operator commanded excessive load."
    ]
    return {"active_hypotheses": hypotheses}

def evidence_validator_node(state: DowntimeState):
    """Acts as a Critic: Checks if hypotheses are supported by evidence."""
    evidence = state["gathered_evidence"]
    hypotheses = state["active_hypotheses"]

    # The Critic LLM evaluates the claims
    # Simulated logic: It finds evidence for H1, but none for H2 or H3
    feedback = "H1 is strongly supported by the overdue CMMS record and historian vibration trend. H2 and H3 lack supporting electrical or MES data. Proceed to draft report focusing on H1."

    # Determine routing based on feedback
    if "Proceed to draft" in feedback:
        return {"validation_feedback": feedback, "route": "draft_report"}
    else:
        # If evidence is weak, route back to gather more data
        return {"validation_feedback": feedback, "route": "gather_more_data"}

def report_writer_node(state: DowntimeState):
    """Synthesizes the final 8D RCA report."""
    # LLM generates the structured report
    report = f"8D Root Cause Analysis for {state['asset_id']}\n"
    report += f"Problem: {state['fault_code']}\n"
    report += f"Root Cause: Bearing failure caused by missed lubrication schedule (overdue by 300 hours).\n"
    report += f"Evidence: Vibration RMS peaked at 15.4 mm/s. CMMS shows missed PM.\n"
    report += f"Corrective Action: Replace bearing. Update CMMS alerting rules to escalate overdue PMs to management."

    return {"final_rca_report": report}
```

```
# 3. Build the Orchestration Graph
workflow = StateGraph(DowntimeState)

workflow.add_node("DataGatherer", data_gatherer_node)
workflow.add_node("HypothesisGenerator", hypothesis_generator_node)
workflow.add_node("EvidenceValidator", evidence_validator_node)
workflow.add_node("ReportWriter", report_writer_node)

workflow.set_entry_point("DataGatherer")
workflow.add_edge("DataGatherer", "HypothesisGenerator")
workflow.add_edge("HypothesisGenerator", "EvidenceValidator")

# Conditional routing based on the Critic's evaluation
def route_validation(state: DowntimeState):
    if state.get("iteration_count", 0) > 3: # Prevent infinite loops
        return "ReportWriter"
    if "route" in state and state["route"] == "gather_more_data":
        return "DataGatherer"
    return "ReportWriter"

workflow.add_conditional_edges("EvidenceValidator", route_validation)
workflow.add_edge("ReportWriter", END)

postmortem_app = workflow.compile()
```

This architecture highlights several critical design patterns for industrial agents. First, the separation of concerns: the `HypothesisGenerator` is optimized for creative, lateral thinking (high temperature), while the `EvidenceValidator` is optimized for strict, logical deduction and grounding (low temperature). This Generator-Critic loop is the primary defense against hallucination. If the generator suggests an electrical short, the critic will demand to see the voltage/current telemetry. If that data is missing from the state, the critic forces the workflow back to the `DataGatherer` node to retrieve it.

Second, tool execution in OT environments requires robust error handling. Industrial databases are often legacy systems hosted on-premises. A SQL query to a 15-year-old CMMS might time out, or the historian API might rate-limit the agent. The `DataGatherer` node must implement idempotency, exponential backoff, and graceful degradation. If the historian is completely unreachable, the agent must not crash; it should note the missing data in its state, proceed with the CMMS and MES data, and explicitly state in the final report that the conclusion is of lower confidence due to the inability to verify time-series telemetry.

1.96 Production Deployment, Trade-offs, and Continuous Learning

Deploying an autonomous diagnostic agent into a live production environment requires strict governance. The output of an RCA dictates physical actions—ordering expensive replacement parts, altering machine parameters, or changing standard operating procedures (SOPs). An AI hallucination here is not merely an inconvenience; it is a safety hazard and a financial liability. Therefore, the system must be deployed with a mandatory Human-in-the-Loop (HITL) architecture.

The agent does not directly execute corrective actions. Instead, it acts as a highly capable analyst, preparing a comprehensive draft of the 8D report. This draft is routed to a human Reliability Engineer via an enterprise dashboard or a collaboration tool like Microsoft Teams or Google Workspace. The human engineer reviews the agent's logic, verifies the cited evidence (which the agent must provide as direct links or embedded charts), and either approves, modifies, or rejects the report. This HITL gate serves a dual purpose: it ensures operational safety, and it generates high-quality, human-verified training data.

When the engineer approves the report, the system executes a "Continuous Learning Flywheel." The finalized, verified RCA is converted into a structured JSON format, embedded using a high-dimensional embedding model (such as Google's `text-embedding-004`), and ingested into the factory's vector database. The next time a similar fault occurs, the agent's RAG tool will retrieve this past RCA. The agent can then reason: *"A similar vibration signature occurred 6 months ago on Asset CNC_02. The root cause was determined to be a missed lubrication schedule. I will prioritize checking the CMMS for missed PMs on the current asset."* Over time, the factory develops an institutional memory that is instantly accessible and perfectly searchable, immune to employee turnover.

When evaluating the trade-offs of this architecture, organizations must consider token costs versus downtime costs. Processing 50 pages of OEM manuals, 100 rows of CMMS data, and summarized historian telemetry might consume 150,000 tokens per investigation. Using a frontier model, this might cost approximately $0.50 to $1.00 per run. In the context of a manufacturing line where downtime costs $10,000 per minute, the API cost is mathematically negligible. The true cost lies in the initial integration—building the secure API wrappers around legacy OT systems so the agent can access them.

Another critical trade-off is the Context Window vs. RAG decision. While modern models boast context windows of 1 to 2 million tokens (capable of ingesting entire libraries of manuals at once), relying solely on massive context windows for every query introduces latency and increases costs linearly. The best practice for the Postmortem Analyst is a hybrid approach: use RAG to retrieve only the relevant 5-10 pages of the OEM manual based on the specific fault code, but use the large context window to ingest the complete, un-chunked alarm history of the machine for the past 24 hours. This ensures the model has deep, continuous context of the immediate event while efficiently referencing external knowledge.

Ultimately, the Downtime Postmortem Analyst represents a shift from reactive maintenance to intelligent, autonomous reliability engineering. By fusing IT and OT data, reasoning from first principles, and maintaining a continuous learning loop, the agentic system ensures that every failure makes the factory smarter, more resilient, and more productive.

Key Takeaways

- **Cross-Domain Fusion is Essential:** True root causes often span physical machine failures (OT) and process/scheduling errors (IT). Agentic systems excel at traversing these domains autonomously.
- **Time-Series Summarization:** Raw, high-frequency sensor data must be processed into statistical or semantic features via specialized tools before being injected into the LLM context window.
- **Generator-Critic Architecture:** To prevent hallucinations in mission-critical environments, the workflow must include a Critic node (Evidence Validator) that forces the Hypothesis Generator to ground its claims in retrieved data.
- **Human-in-the-Loop (HITL) Safety:** Agents should draft the RCA and recommend actions, but a human engineer must approve the final report before physical changes or purchases are executed.
- **The Flywheel Effect:** Approved RCA reports must be embedded into a vector database, allowing the agent to learn from past failures and instantly recall historical solutions during future downtime events.

Learning Objectives

- Understand the architectural shift from passive, descriptive digital twins to prescriptive, agent-orchestrated simulation environments.
- Design a LangGraph-based orchestration loop capable of translating natural language business intents into deterministic simulation parameters.
- Implement robust tool-calling patterns to interface Large Language Models (LLMs) with external Discrete Event Simulation (DES) engines and physics solvers.
- Evaluate the trade-offs between token consumption, simulation compute latency, and the fidelity of agent-generated mitigation strategies in production environments.
- Apply the Generator-Critic pattern to validate simulation outputs and prevent physical hallucinations in manufacturing scenarios.

1.97 The Evolution from Passive Twins to Agentic Simulators

The concept of the Digital Twin has been a cornerstone of Industry 4.0 for over a decade. Defined formally by standards such as ISO 23247, a digital twin is a virtual representation of a physical asset, process, or system, synchronized at a specified frequency and fidelity. Historically, the maturity of digital twins has been categorized into distinct levels. Level 1 and Level 2 twins are primarily descriptive and diagnostic; they aggregate telemetry from Programmable Logic Controllers (PLCs), Supervisory Control and Data Acquisition (SCADA) systems, and Industrial Internet of Things (IIoT) sensors to populate 3D dashboards. They answer the questions: *"What is happening right now?"* and *"Why did it happen?"*

However, the true value of a digital twin lies in its predictive and prescriptive capabilities—answering *"What will happen if...?"* and *"What should we do about it?"* Traditionally, unlocking this Level 4 and Level 5 maturity required integrating the digital twin with Discrete Event Simulation (DES) software, such as Siemens Plant Simulation, AnyLogic, or FlexSim. While these simulation engines are mathematically rigorous and highly accurate, they suffer from a severe operational bottleneck: the human interface.

Setting up a "what-if" scenario in a traditional DES environment is a highly manual, labor-intensive process. If a supply chain manager wants to know the impact of a three-day delay from a tier-two supplier, an industrial engineer must manually open the simulation model, locate the specific variables representing that supplier's lead time, adjust the statistical distributions, run the simulation, export the results to Excel, analyze the key performance indicators (KPIs) like throughput and queue lengths, and finally draft a report. This process can take hours or even days. By the time the analysis is complete, the window of opportunity to mitigate the disruption may have already closed. The factory operates in real-time, but the simulation operates in batch-time.

This is the exact friction point where Agentic AI introduces a paradigm shift. A Factory Twin Scenario Simulator is not a replacement for the underlying physics or discrete-event math; rather, it is an intelligent, autonomous orchestration layer that sits on top of the simulation engine. By leveraging the reasoning, planning, and tool-use capabilities of frontier models like Gemini 3.5 Pro, the agentic simulator bridges the semantic gap between high-level business intent and low-level simulation parameters.

When a production manager asks the agent, *"Simulate the impact of a 15% yield drop on the SMT (Surface Mount Technology) line and propose a way to maintain our weekly quota,"* the agentic system performs a complex sequence of cognitive and executable tasks. First, it parses the natural language to identify the target asset (SMT line), the perturbed variable (yield rate), the magnitude of the perturbation (-15%), and the objective function (maintain weekly quota). Second, it queries the factory's semantic knowledge graph to map "SMT line yield" to the exact JSON payload required by the simulation API (e.g., `{"asset_id": "SMT_01", "parameter": "defect_rate", "value": 0.15}`).

Crucially, the agent does not just run one simulation. Because it operates at machine speed, it can design a Design of Experiments (DoE). To achieve the objective of maintaining the weekly quota, the agent might autonomously hypothesize three different mitigation strategies: (A) increasing the conveyor speed by 5%, (B) scheduling a weekend overtime shift, or (C) rerouting partial sub-assemblies to a legacy line. It translates all three hypotheses into simulation parameters, triggers the DES engine via API calls concurrently, waits for the asynchronous results, and then acts as a data analyst. It compares the simulated KPIs—evaluating the trade-offs between the energy cost of running faster, the labor cost of overtime, and the quality risks of the legacy line. Finally, it synthesizes these findings into a concise, actionable recommendation for the production manager.

This architecture fundamentally changes the economics of scenario planning. The marginal cost of exploring a new operational hypothesis drops from hours of engineering time to mere cents in API token costs. Furthermore, it democratizes access to the digital twin. Floor supervisors, supply chain analysts, and maintenance technicians can interact with the factory's mathematical model using natural language, without needing to learn proprietary simulation software.

However, delegating simulation orchestration to an LLM introduces unique risks. LLMs are probabilistic text generators; they do not inherently understand the laws of physics, thermodynamics, or factory kinematics. If left unconstrained, an LLM might propose a mitigation strategy that violates physical reality—such as increasing a machine's throughput to 500% of its rated capacity, or moving inventory instantly without accounting for automated guided vehicle (AGV) transit times. Therefore, the architecture of the Factory Twin Scenario Simulator must be heavily constrained by deterministic guardrails, strict API schemas, and a robust Generator-Critic loop that validates every proposed parameter against the physical limits of the factory before the simulation is even executed.

1.98 Architecting the Agentic Simulation Loop

To build a production-grade Factory Twin Scenario Simulator, we must move beyond simple prompt-response paradigms and implement a stateful, cyclical cognitive architecture. The agent must be able to plan, execute, observe the results, and iteratively refine its approach if the initial simulation fails to meet the objective function. Frameworks like LangGraph or the Google Agent Development Kit (ADK) are ideal for this, as they allow us to define the reasoning process as a directed graph with explicit state management.

The architecture consists of four primary cognitive nodes, orchestrated around a central state object. The state object maintains the history of the user's request, the current simulation parameters, the raw outputs from the simulation engine, and the agent's internal critique.

The four primary nodes are:

1. **Intent Parser & Planner:** This node receives the natural language query. Its job is to extract the core entities (assets, timeframes, variables) and define the objective function. It outputs a structured plan of which scenarios need to be simulated.
2. **Parameter Mapper (Generator):** This node translates the high-level plan into the exact JSON schema required by the simulation API. It queries a vector database containing the factory's semantic ontology to ensure it uses the correct variable names (e.g., mapping "conveyor speed" to `var_conv_vel_m_s`).
3. **Physical Bounds Critic:** Before any API call is made, this deterministic (or low-temperature LLM) node checks the proposed JSON payload against a strict schema of physical constraints. If the Generator proposes a conveyor speed of 10 m/s, but the physical limit is 2 m/s, the Critic rejects the payload and forces the Generator to try again. This is the primary defense against physical hallucinations.
4. **Simulation Executor & Analyst:** This node handles the asynchronous API calls to the DES engine (e.g., via a REST API to an Azure Digital Twin or a headless AnyLogic model). Once the simulation completes, it fetches the results. Because simulation logs can be massive (gigabytes of time-series data), this node uses deterministic Python tools (like Pandas) to extract summary KPIs (mean throughput, max queue length) rather than feeding raw logs into the LLM context window. Finally, it synthesizes the KPIs into a human-readable recommendation.

Below is a production-ready implementation of this orchestration loop using Python and LangGraph. Notice how the state is explicitly typed, and how the conditional routing enforces the Generator-Critic pattern.

Listing 1.51: LangGraph Implementation of the Factory Twin Simulator

```
import json
```

```python
from typing import Annotated, List, Dict, Any, Optional
from typing_extensions import TypedDict
from langgraph.graph import StateGraph, END
from langchain_core.messages import HumanMessage, SystemMessage

# 1. Define the Simulation State
class SimulationState(TypedDict):
    user_query: str
    objective: str
    proposed_parameters: List[Dict[str, Any]]
    validation_status: str
    simulation_results: List[Dict[str, Any]]
    final_recommendation: str
    iteration_count: int

# 2. Define the Nodes
def intent_parser_node(state: SimulationState):
    """Extracts the objective and identifies variables to perturb."""
    # In production, this calls an LLM with structured output (JSON mode)
    # Simulated output for: "What if SMT yield drops 15%? How do we hit quota?"
    return {
        "objective": "Maintain weekly quota of 10,000 units despite yield drop.",
        "iteration_count": state.get("iteration_count", 0) + 1
    }

def parameter_mapper_node(state: SimulationState):
    """Generates the JSON payload for the simulation API."""
    # LLM generates hypotheses to solve the objective
    # Hypothesis 1: Increase speed. Hypothesis 2: Add overtime.
    params = [
        {"scenario": "Base + Yield Drop", "smt_yield": 0.85, "line_speed": 1.0, "shifts": 10},
        {"scenario": "Mitigation A: Speed Up", "smt_yield": 0.85, "line_speed": 1.2, "shifts": 10},
        {"scenario": "Mitigation B: Overtime", "smt_yield": 0.85, "line_speed": 1.0, "shifts": 12}
    ]
    return {"proposed_parameters": params}

def physical_critic_node(state: SimulationState):
    """Validates parameters against physical factory constraints."""
    # Deterministic check against a known schema
    MAX_SPEED = 1.15 # Factory limit is 115% of nominal

    for param in state["proposed_parameters"]:
        if param.get("line_speed", 1.0) > MAX_SPEED:
            feedback = f"REJECTED: line_speed {param['line_speed']} exceeds physical limit of {MAX_SPEED}."
            return {"validation_status": feedback}

    return {"validation_status": "APPROVED"}
```

```python
def simulation_executor_node(state: SimulationState):
    """Executes the API calls to the DES engine and extracts KPIs."""
    # Simulated API call to headless simulation engine
    results = []
    for param in state["proposed_parameters"]:
        # Mocking the simulation math
        base_output = 10000
        actual_output = base_output * param["smt_yield"] * param["line_speed"] * (param["shifts"]/10)
        results.append({
            "scenario": param["scenario"],
            "simulated_output": int(actual_output),
            "cost_delta": (param["line_speed"] - 1.0)*5000 + (param["shifts"] - 10)*2000
        })
    return {"simulation_results": results}

def result_analyst_node(state: SimulationState):
    """Synthesizes the KPIs into a final recommendation."""
    # LLM analyzes the results array and drafts the response
    res = state["simulation_results"]
    rec = f"Analysis complete. To maintain the 10,000 unit quota with an 85% yield:\n"
    rec += f"- Mitigation A (Speed Up) was rejected by physical constraints.\n"
    rec += f"- Mitigation B (Overtime) achieves {res[2]['simulated_output']} units at an additional cost of ${res[2]['cost_delta']}.\n"
    rec += "Recommendation: Approve Mitigation B (schedule 2 extra shifts)."
    return {"final_recommendation": rec}

# 3. Build the Orchestration Graph
workflow = StateGraph(SimulationState)

workflow.add_node("IntentParser", intent_parser_node)
workflow.add_node("ParameterMapper", parameter_mapper_node)
workflow.add_node("PhysicalCritic", physical_critic_node)
workflow.add_node("SimulationExecutor", simulation_executor_node)
workflow.add_node("ResultAnalyst", result_analyst_node)

workflow.set_entry_point("IntentParser")
workflow.add_edge("IntentParser", "ParameterMapper")
workflow.add_edge("ParameterMapper", "PhysicalCritic")

# Conditional routing based on the Critic's evaluation
def route_validation(state: SimulationState):
    if state.get("iteration_count", 0) > 3:
        return "ResultAnalyst" # Fallback to prevent infinite loops
    if "REJECTED" in state["validation_status"]:
        return "ParameterMapper" # Force LLM to try again with feedback
    return "SimulationExecutor"

workflow.add_conditional_edges("PhysicalCritic", route_validation)
workflow.add_edge("SimulationExecutor", "ResultAnalyst")
workflow.add_edge("ResultAnalyst", END)
```

```

twin_app = workflow.compile()
```

This code illustrates a critical design pattern for industrial agents: the LLM is never allowed to directly execute a tool that impacts the physical world (or in this case, the heavy-compute simulation environment) without passing through a deterministic validation gate. The `PhysicalCritic` node acts as a hard boundary. If the LLM hallucinates a line speed of 1.2 (120%), the critic catches it, rejects the state, and routes the graph back to the `ParameterMapper` with the explicit error message injected into the prompt context. The LLM then "learns" from this rejection and generates a new, compliant parameter set.

Furthermore, the `SimulationExecutor` node highlights the necessity of asynchronous tool calling. A high-fidelity discrete event simulation of a factory floor might take 5 to 15 minutes to compute. An agentic system cannot hold an HTTP connection open for that long. In a true production deployment, the `SimulationExecutor` would submit a job to a queue (e.g., via Google Cloud Pub/Sub or AWS SQS), transition the LangGraph state to a "waiting" status, and hibernate. A separate webhook or polling mechanism would wake the agent up once the simulation engine returns the payload, allowing the graph to resume execution at the `ResultAnalyst` node. This event-driven architecture is essential for scaling agentic twins across enterprise operations.

1.99 Real-World Scenarios and Trade-off Analysis

Deploying a Factory Twin Scenario Simulator transforms how organizations handle volatility, uncertainty, complexity, and ambiguity (VUCA) in their operations. By automating the translation of intent to simulation, the agent enables rapid, data-backed decision-making in scenarios where human intuition often fails due to the non-linear dynamics of manufacturing systems.

Consider a real-world scenario: Dynamic Line Balancing during a machine failure. In a highly automated machining cell, if CNC Machine 3 experiences a catastrophic spindle failure, the immediate human reaction is often to route all pending work to CNC Machine 4. However, a factory is a complex network of queues. Rerouting all work to Machine 4 might cause a massive bottleneck, starving downstream assembly stations and ultimately reducing overall factory throughput more than if the work was simply paused.

When this failure event is detected by the agent (via an alert from the CMMS or IoT platform), it automatically spins up the digital twin. It formulates multiple hypotheses and runs each through the DES engine; Table~1.7 summarizes the simulated outcomes. The agent presents the counter-intuitive, mathematically proven recommendation—Scenario B—to the shift supervisor within minutes of the initial failure.

While the operational benefits are immense, architects must carefully manage the trade-offs inherent in this system, specifically regarding cost, latency, and the "Sim2Real" (Simulation-to-Reality) gap.

Token Costs vs. Compute Costs: In an agentic simulator, there are two distinct cost centers. The first is the LLM token cost for reasoning, planning, and analyzing. The second is the cloud compute cost for running the physics/DES engine. Interestingly, the token costs are usually

Table 1.7: Simulated mitigation scenarios for the CNC 3 spindle failure

Dimension	Scenario A	Scenario B	Scenario C
Mitigation action	Reroute all work to CNC 4	Use existing buffer stock while awaiting standard maintenance	Pay a premium to expedite maintenance while running CNC 4 at 110% speed
Simulated outcome	Downstream starvation event costing $10,000 in lost production	Buffer sustains the line for exactly 4 hours—enough for standard repair with no expedite fees or bottlenecks (recommended)	$5,000 in expedite fees

negligible compared to the compute costs. A complex reasoning trace using Gemini 3.5 Pro might consume 20,000 tokens, costing a few cents. However, spinning up a GPU-accelerated NVIDIA Omniverse instance or a heavy AnyLogic model in the cloud might cost several dollars per run. Therefore, the agent must be designed to be computationally frugal. It should not run 1,000 Monte Carlo simulations blindly. Instead, it should use Bayesian optimization or LLM-guided heuristics to narrow down the search space to 3-5 highly probable scenarios before invoking the expensive simulation tool.

Latency and Asynchronous UX: As discussed, simulations take time. If a user asks a chatbot a question and has to wait 15 minutes for a response, the user experience is broken. The UI/UX for an Agentic Twin must be asynchronous. When the user submits a prompt, the agent should immediately reply with its plan: *"I understand you want to evaluate the SMT yield drop. I have formulated 3 mitigation scenarios and dispatched them to the simulation cluster. I will notify you when the results are ready."* This requires integrating the agent with enterprise notification systems (e.g., Slack, Microsoft Teams, or email) rather than relying solely on a synchronous chat window.

The Sim2Real Gap and Hallucinations: The most dangerous failure mode of this system is not an LLM hallucination, but a simulation hallucination—where the digital twin's model of the factory has drifted from physical reality. If the simulation assumes a conveyor moves at 2 m/s, but a mechanical issue has degraded it to 1.5 m/s in the real world, the agent's recommendations will be mathematically correct within the simulation but disastrous on the floor.

To mitigate the Sim2Real gap, the agentic architecture must include a continuous calibration loop. Before running a predictive scenario, the agent should first run a "diagnostic" simulation using the exact parameters from the last 24 hours of actual factory telemetry. It then compares the simulated output against the actual historical output. If the delta exceeds a predefined threshold (e.g., 5%), the agent must flag the digital twin as "out of calibration" and refuse to provide predictive recommendations until an engineer updates the underlying physics model. This self-awareness—the ability of the agent to know when its tools are untrustworthy—is the hallmark of a mature, enterprise-grade autonomous system.

Ultimately, the Factory Twin Scenario Simulator represents the convergence of generative AI and deterministic physics. By wrapping rigorous mathematical models in an accessible, reasoning-capable agentic layer, manufacturers can transition from reacting to disruptions to proactively engineering resilience into their daily operations.

Key Takeaways

- **Agents as Orchestrators, Not Solvers:** The LLM should never attempt to calculate factory physics or queueing theory itself. Its role is to orchestrate external, deterministic Discrete Event Simulation (DES) engines via structured API calls.
- **Mandatory Generator-Critic Loops:** Always implement a deterministic Critic node to validate proposed simulation parameters against physical factory constraints before executing expensive compute jobs.
- **Asynchronous Execution is Required:** Because high-fidelity simulations take minutes to hours, the agent architecture and the user interface must support asynchronous, event-driven workflows (e.g., pub/sub queues and webhook callbacks).
- **Data Extraction over Raw Ingestion:** Never feed raw simulation logs (gigabytes of time-series data) back into the LLM context window. Use deterministic tools (Python/Pandas) to extract summary KPIs for the agent to analyze.
- **Guard against the Sim2Real Gap:** Implement automated calibration checks where the agent compares recent simulation outputs against actual historical telemetry to ensure the digital twin remains an accurate reflection of reality.

www.ingramcontent.com/pod-product-compliance
Lightning Source LLC
LaVergne TN
LVHW070851210926
848482LV00016B/910